TECHNOLOGY TRANSFER AGREEMENTS
AND THE EC COMPETITION RULES

Technology Transfer Agreements and the EC Competition Rules

VALENTINE KORAH

CLARENDON PRESS · OXFORD
1996

Oxford University Press, Great Clarendon Street, Oxford OX2 6DP

Oxford New York
Athens Auckland Bangkok Bombay
Calcutta Cape Town Dar es Salaam Delhi
Florence Hong Kong Istanbul Karachi
Kuala Lumpur Madras Madrid Melbourne
Mexico City Nairobi Paris Singapore
Taipei Tokyo Toronto
and associated companies in
Berlin Ibadan

Oxford is a trade mark of Oxford University Press

Published in the United States
by Oxford University Press Inc., New York

British Library Cataloguing in Publication Data
Data available

Library of Congress Cataloging in Publication Data
Korah, Valentine.
Technology transfer agreements and the EC competition rules.
p. cm.
1. Technology transfer—Law and legislation—European Union
countries.   2. Licence agreements—European Union countries.
3. Restraint of trade—European Union countries.   I. Title.
KJE2777.K67 1996
341.7'59—dc20   96-35168
ISBN 0–19–826243–4

1 3 5 7 9 10 8 6 4 2

Typeset by Cambrian Typesetters, Frimley, Surrey
Printed in Great Britain on acid-free paper by
Biddles Ltd., Guildford and King's Lynn

Preface

Encouraged by the favourable reception of and continuing demand for my five commentaries on group exemptions by the EC Commission, I have decided to add a monograph on the technology transfer regulation.

In accordance with my earlier practice, I have tried to analyse the legislation in the context of the case law of both the Commission and the Court. I still believe that many agreements that the Commission has exempted should have been cleared, and would rely on this view if it were necessary when seeking to enforce a contract in a national court. Indeed there are signs that many officials in DG IV now accept this view. Nevertheless, cautious draftsmen will prefer to play safe and bring their licensing agreements within the four corners of the regulation. Doubtless the regulation will be revised on its expiry after ten years. So I have not refrained from criticism. The Community Court may well look to policy considerations when it comes to rule on its interpretation and national courts should do the same. I hope that the monograph may be a useful working tool for business advisers.

The monograph starts by describing the Commission's bifurcation of article 85(1) and (3), and the two concepts adopted by the Court to find that an agreement does not infringe article 85(1), so needs no exemption; the doctrine of ancillary restraints and the need to analyse an agreement in its economic context before finding it anti-competitive. I have then taken advantage of the recent adoption by the Department of Justice and the Federal Trade Commission in the United States of new guidelines on intellectual property licensing to describe them in general and, as I come to various particular clauses, I have described the American analysis of the policy considerations as a useful background to the European.

After considering the general extent of the regulation and the specific exclusions and inclusions, I have analysed its substantive provisions clause by clause and appended selected articles of the EC and EEA Treaties and the whole regulation. The latter I have annotated to help readers find the relevant recitals when reading the articles and so forth.

I would like to thank many people for help with this book. Signor Enrique Gonzalez Diaz, who works for the Legal Service of the Commission, spent part of his holiday reading it and made many helpful criticisms of specific points. I am indebted to support both from Fordham University School of Law, where I started to write this monograph shortly before the adoption of the regulation and to University College London, where I finished it in the early summer. I have visited Fordham for six spring semesters in a row, and am always generously treated: provided with an office and a fast computer, as well as funding for research assistance and the pleasure and stimulation of interesting colleagues. UCL has kept on a retired Professor as a part-time member of the staff, blessed with an office and various forms of support including further funding for a research assistant.

At Fordham, Zain Husain checked many citations of cases and articles, despite his many other activities as a third-year student, articles editor to *Fordham International Law Journal* and a candidate for the New York Bar examinations. He was unable to finish and I am most grateful to Chris Loweth, a graduate of UCL, who worked intensively on the book at the final stages between completing the course for the English Bar and pupillage. He has been constructive, loyal, and flexible. He also checked many citations, ensured that I used the OUP house style and saved me from many infelicities and errors.

I would also like to thank Richard Hart who enthusiastically accepted my offer to write this book for OUP, although, until recently, it has been better known in the academic than the professional lawyers' market. He also supported some research assistance and readily agreed to let me have my favourite copy editor, Kate Elliott. She takes great pains to make tables of cases and legislation as useful as possible, and has developed excellent ways of providing a great deal of information simply. She is responsible for preparing the table of legislation, although, with the help of my research assistants, I have on this occasion prepared the table of cases.

Elissa Soave of OUP has been very constructive and efficient in solving problems and the typesetters, Cambrian, worked fast and efficiently, producing from disk most of the page proofs in only about two weeks. Without the help of all these people and institutions, this book would not have been so well prepared. I hasten to add that there must remain many errors, and for these only I am responsible.

The book was finished on 2 August 2 1996, less than six months after the regulation was published.

Valentine Korah

Guildford Chambers,
Stoke House,
Leapale Lane,
Guildford, England,
GU1 4LI,
tel 00 44 (1483) 391 31
fax 00 44 (1483) 300 542

Contents

Legislation

Treaties and Conventions

1883 Paris. Union Convention for the Protection of Industrial Property (20 March), (Stockholm Revision, 14 July 1967), 828 UNTS 11851   5.2.4
1957. Rome. Treaty establishing the European Community (as amended) (EC Treaty)

Art. 2	**App. 1**
3	**App. 1**
(g)	2.1.3
5	**App. 1**
(2)	2.1.3
Arts. 30–36	2.1
Art. 30	2.1, 2.1.2, 2.1.4, 7.3.14, **App. 1**
36	2.1, 2.1.1, 2.1.2, 2.1.3, 7.3.14, **App. 1**
85	1.1.1, 1.2, 2.1, 2.1.1, 2.1.2, 2.1.4, 3.1.8, 3.2.3, 4.2, 4.2.1, 4.2.2, 4.2.3, 5.1.1.2, 5.1.1.3, 7.3.5, 7.3.6.3, 9.1.1, 9.1.2, **App. 1**
(1)	1.1, 1.1.1, 1.1.2, 1.2, 1.3.1, 1.4.1, 1.4.2, 1.4.2.1, 1.4.3, 2.1, 2.1.1, 2.1.2, 2.1.3, 2.2, 2.2.1, 2.2.3, 2.2.4, 2.2.5, 3.1.2, 3.1.3, 3.1.4, 3.1.5, 3.1.7, 3.1.8, 3.2.1, 3.2.5, 3.4.1, 3.4.2, 4.1.3, 4.1.4, 4.2, 5.1, 5.1.1.1, 5.1.1.1.2, 5.1.1.1.3, 5.1.1.2, 5.1.2.2, 5.1.4, 5.2.2, 5.2.3, 6.6, 6.6.4, 6.6.7, 7.1, 7.2, 7.3.1, 7.3.2, 7.3.3, 7.3.4.3, 7.3.4.4, 7.3.5.2, 7.3.6.1, 7.3.6.2, 7.3.6.3, 7.3.9.2, 7.3.13, 7.3.17, 8.2.2, 8.2.3, 8.5.1, 8.5.2, 8.5.3, 8.6, 9.1.1, 9.1.3.4, 9.2.2, Ch. 10
(2)	1.1, 1.1.2
(3)	1.1, 1.1.1, 1.4.2, 1.4.2.1, 3.1.2, 3.1.5, 6.6.5, 6.9, 8.1, 8.2.2, 8.3.1, 8.5.1, 8.5.3, 9.1, 9.1.1, Ch. 10
86	1.2, 1.4.1, 2.1.1, 2.1.2, 2.3, 4.2.3, 7.3.5, 8.5.3, 9.1.2, **App. 1**
87	8.4
89	8.4
173	1.4.1, 8.5.4, 8.5.4.1, 8.5.4.4, 8.5.4.5
175	8.5.4, 8.5.4.4, 8.5.4.5
77	1.4.1, 5.1.1.1.2, 7.3.6.3, 8.5.4.1, 8.5.4.5
189(b)	3.3
190	1.1.1, 8.5.4, 9.1.1

Regulations

Reg. 556/89, [1989] OJ L461/1 (group exemption for know-how licensing
agreements) corr. [1990] OJ L257/15, [1989] 4 CMLR 774

Directives

Notices

National Legislation

United Kingdom

United States

Cases

The table includes not only the formal decisions of the Commission, but also its informal ones described in an annual report, or a notice in the Official Journal under article 19(3) of regulation 17. The bold numbers indicate where a case is discussed.

E.C. Commission, the Court of First Instance, and the European Court of Justice

Alphabetical Table

AEG-Telfunken—Allgemeine Elektricitäts-Gesellschaft AEG Telfunken AG *v.* Commission (107/82), 25 October 1983, [1983] ECR 3151, [1984] 3 CMLR 325, CMR 14,018.
   7.3.17

Ahmed Saeed Flugreisen *v.* Zentrale zur Bekämpfung unlauteren Wettbewerbs (66/86), 11 April 1989, [1989] ECR 803, [1990] 4 CMLR 102, [1989] 2 CEC 654.
   8.1, 8.3.1, 8.5.1, **9.1.2**

AKZO Chemie BV and AKZO Chemie UK Ltd *v.* Commission (53/85), 24 June 1986, [1986] ECR 1965, [1987] 1 CMLR 231, CMR 14,318.
   6.1.1.3, 7.3.1, 8.2.4, 8.5.4, 8.5.4.1

Allen & Hanburys Ltd. *v.* Generics (UK) Ltd (434/85), 3 March 1988, [1988] ECR 1245, [1988] CMLR 701, CMR 14,446.
   2.1.3

AOIP/Beyrard (76/29/EEC), 2 December 1975, [1976] OJ L6/8, [1976] 1 CMLR D14, CMR 9801.
   1.3.1, **2.2**, 3.2.1, 7.3.6.3, 7.3.7

Association of Plant Breeders of the EEC (Comasso) (IV/31/318), 11 January 1990, Art. 19(3) Notice, [1990] OJ C6/3, [1990] 4 CMLR 259.
   1.4.1

Automec II—Automec Srl *v.* Commission (T-24/90), 18 September 1992, [1992] ECR II–2223, [1992] 5 CMLR 431.
   1.1.1, 8.5.4.4

IGR Stereo Television, *Eleventh Report on Competition Policy*, point 63, *Fourteenth Report on Competition Policy*, point 76.
    5.1.1.1.2, 5.1.1.1.3

Italian Flat Glass (89/93/EEC), 7 December 1988, [1989] OJ L33/44, [1990] 4 CMLR 535, [1989] 1 CEC 2077.
    9.1.2

Italian Flat Glass—Società Italiana Vetro SpA *v.* Commission (T–68, 77, & 78/89), 10 March 1992, [1992] ECR II–1403, [1992] 5 CMLR 302, [1992] 2 CEC 33.
    9.1.2

Italy *v.* Council and Commission (32/65), 13 July 1966, [1966] ECR 389, [1969] CMLR 39, CMR 8048.
    8.5.1

Junghans GmbH (Re the Agreement of Gebrüder) (77/100/EEC), 21 December 1976, [1977] OJ L30/10, [1977] 1 CMLR D82, CMR 9912.
    4.2, 5.1.1.3, 5.1.2.5, 6.9

KSB/Lowara/Goulds/ITT (91/38/EEC), 12 December 1990, [1991] OJ L19/25, [1992] 5 CMLR 55, [1991] 1 CEC 2009.
    5.1.1.2

Kabelmetal/Luchaire (75/494/EEC), 18 July 1975, [1975] OJ L222/34, [1975] 2 CMLR D40, CMR 9761.
    1.1, 2.2, 7.3.1, 7.3.10

Keurkoop BV *v.* Nancy Kean Gifts BV (144/81), 14 September 1982, [1982] ECR 2853, [1983] 2 CMLR 47, CMR 8861.
    2.1.2, 2.1.3, 3.3

La Technique Minière—see Société La Technique Minière.

Lancôme—SA Lancôme and Cosparfrance Nederland BV *v.* Etos BV and Albert Heijn Supermart BV (99/79), 10 July 1980, [1980] ECR 2511, [1981] 2 CMLR 164, CMR 8714.
    1.1.1

Langnese-Iglo GmbH & Co. KG (93/406/EC), 23 December 1991, [1993] OJ L183/19, [1994] 4 CMLR 51, [1993] 2 CEC 2123.
    3.4.2, 6.7.1, **9.1.1**

Langnese-Iglo GmbH & Co. KG *v.* Commission (T–7/93), 8 June 1995, [1995] 5 CMLR 602, [1995] 2 CEC 217.
    9.1.1

Spices: Brooke Bond Liebig (78/172/EEC), 21 December 1977, [1978] OJ L53/20, [1978] 2 CMLR 116, CMR 10,017.
  6.9

Standard Seed Production and Sales Agency in France (IV/31.318), 11 January 1990, [1990] OJ C6/3, [1990] 4 CMLR 259.
  5.2.4

Stergios Delimitis *v.* Henniger Bräu—see Delimitis (Stergios) *v.* Henniger Bräu AG.

Sugar Cases—Re the European Sugar Cartel: Cooperatiëve Vereniging 'Suiker Unie' UA *v.* Commission (40–48, 50, 54–56, 111, & 113–114/73), 16 December 1975, [1975] ECR 1663, [1976] 1 CMLR 295, CMR 8334.
  8.2.2, 8.4

Suralmo, *Ninth Report on Competition Policy*, at point 72.
  7.3.2

Tetra Pak I (BTG Licence) (88/501/EEC), 26 July 1988, [1988] OJ L272/27, [1990] 4 CMLR 47.
  6.7.1, **9.1.1**, **9.1.2**, 9.1.3.1
and, on appeal,
Tetra Pak Rausing SA *v.* Commission (T–51/89), 10 July 1990, [1990] ECR II–309, [1991] 4 CMLR 334, [1990] 2 CEC 409.
  9.1.1, 9.1.2

Thetford Corporation *v.* Fiamma SpA (35/87), 30 June 1988, [1988] ECR 3585, [1988] 3 CMLR 549, CMR 14,497.
  2.1.3, 2.2.5

Transocean Marine Paint Association (88/635/EEC), 2 December 1988, [1988] OJ L351/40, [1989] 4 CMLR 621, 1 CEC 2003.
  3.1.6, 5.1.1.1.2

VAG France SA *v.* Etablissements Magne SA (10/86), 18 December 1986, [1988] ECR 4071, [1988] 4 CMLR 98, CMR 14,390.
  3.4.1

Vaessen/Moris (79/86/EEC), 10 January 1989, [1979] OJ L19/32, [1979] 1 CMLR 511, CMR 10,107.
  7.3.5.2

Velcro/Aplix (85/410/EEC), 12 July 1985, [1985] OJ L233/22, [1989] 4 CMLR 157, CMR 10,715.
  3.2.2, **7.2**, 7.3.3, 7.3.4.3, 7.3.4.4, **7.3.5.2**

Viho Europe BV *v.* Commission (T–102/92), 12 January 1995, [1995] ECR

National cases

England and Wales

1

Introduction

1.1 BIFURCATION OF ARTICLE 85(1) AND (3) OF THE EC TREATY

As is well known, article 85(1) of the EC Treaty prohibits as incompatible with the Common Market collusion between undertakings that may affect trade between Member States and that has the object or effect of restricting, preventing, or distorting competition within the Common Market. Article 85(3) provides for exemptions and article 85(2) provides that agreements that infringe the article as a whole are void *pro tanto*.[1]

In many individual decisions relating to agreements notified to it, the Commission of the European Communities, which has exclusive power to exempt anti-competitive agreements under article 85(3), has maintained

---

[1] Art. 85 is reproduced in App. 1.

When considering whether trade between Member States may be affected, one should look to the agreement as a whole, and not merely to the particular provisions that may restrain competition: *Windsurfing International Inc.* v. *Commission* (193/83), [1986] ECR 611, [1986] 3 CMLR 489, CMR 14,271, para. 96. 'Trade' is widely interpreted to include industry, services, and freedom to establish a business in another Member State. *Pronuptia de Paris GmbH* v. *Pronuptia de Paris Irmgard Schillgalis* (161/84), [1986] ECR 353, [1986] 1 CMLR 414, CMR 14,245, para. 26. The ruling probably includes all the principles of free movement.

When deciding whether an agreement has the object or effect of restricting competition, and how much of the contract is void, one should look to particular provisions, to the agreement as a whole, and to the economic context of the agreement: *Société La Technique Minière* v. *Maschinenbau Ulm GmbH* (56/65), [1966] ECR 235, [1966] CMLR 357, CMR 8047, and *Delimitis (Stergios)* v. *Henniger Bräu* AG (C–234/89), [1991] ECR I–935, [1992] 5 CMLR 210, [1992] 2 CEC 530. This may involve considering whether similar vertical agreements are widespread.

The Court has also developed a concept of ancillary restraints necessary to make viable a transaction which does not in itself have the object or effect of restricting competition. In *Pronuptia*, for instance, the Court looked first to the general category of distribution franchising and found that, in itself, it did not restrict competition contrary to art. 85(1). After considering a number of individual clauses which it cleared as being necessary to make franchising viable, the Court assessed two clauses in combination: (1) the obligation of each franchisee to sell only from the particular shop; and (2) an exclusive territory. These seem to have given each franchisee virtually absolute territorial protection from the franchisor and other franchisees and were found to infringe art. 85(1) once the network was widespread. The Court referred also to the economic context of the agreement when it summarized what it had said earlier.

Only the provisions that have the object or effect of restricting competition are void by virtue of Community law. It is for national law to decide whether the rest can stand: *Société La Technique Minière* and *Société de Vente de Ciments et Bétons de l'Est SA* v. *Kerpen & Kerpen* (319/82), [1983] ECR 4173, [1985] 1 CMLR 511, CMR 14,043.

For the extent of invalidity under English law, see 1.1.2 below.

control over the application of the article. It frequently finds that a restraint of conduct which has significant effects on the market has the object or effect of restricting competition, and then exempts the agreement on the ground that without that restriction the market would, in effect, be less competitive. Any restraint on parallel trade is likely to be so treated, even if it be required to induce investment.

In decisions concerning patent licences, such as *Davidson Rubber*[2] and *Kabelmetal*,[3] the Commission held that the grant of a network of exclusive territories and the ancillary restraints on the licensees producing or selling outside their territories restricted competition, in that the patentee gave up its freedom to grant a licence for the territory to someone else who might have exported to another Member State. The Commission went on to exempt an exclusive manufacturing licence on the ground that, without such protection, the technology would not have been available in Europe.[4] In that case, it is hard to see what competition was actually restricted.

The Commission's finding that the restriction of conduct—the promise not to license anyone else in the territory—restricted competition was purely formalistic, since it found that such conduct was not commercially practicable. It was only under article 85(3) that it looked realistically at the market. In effect, there was a *per se* prohibition of exclusive technology licences, subject only to a *de minimis* rule, applied not when the licence was granted but when the Commission's decision was adopted.[5]

As recently as the end of 1988, the Commission was still refusing to clear exclusive licences, but exempting them, even when the parties did not compete with each other before the licence[6] and the exclusivity related only to manufacture, all the licensees being allowed to sell throughout the common market. When granting an exemption in *Rich Products/Jus-Rol*,[7] the Commission stated that an exclusive licence of know-how for

---

[2] [1972] OJ L143/31, [1972] CMLR D52, CMR 9512, paras. 35–40.

[3] [1975] OJ L222/34, [1975] 2 CMLR D40, CMR 9761, paras. 25–8.

[4] *Davidson Rubber*, n. 2 above, paras. 47–8; *Kabelmetal*, n. 3 above, paras. 36–41. The Commission does not spell out its reasoning but, as explained shortly at 1.4.2 below, a licensee has to invest in tooling up and making the product known in its territory before it can make any money. Before this expense has been paid off, it will have to start paying royalties and, often, a payment in advance. To induce this investment, the licensee needs to be assured that it will be able to appropriate the benefit. An exclusive licence with export prohibitions on other licensees may be an appropriate way for the innovator to encourage each licensee to make the commitments.

[5] *Burroughs/Delplanque* [1972] OJ L13/50, [1972] CMLR D72, CMR 9485, para. 6; *Burroughs/Geha* [1972] OJ L13/53, [1972] CMLR D72, CMR 9485; and *Davidson Rubber* n. 2 above. In only two decisions relating to exclusive licences that it found infringed art. 85(1) did the Commission not grant an individual exemption after forcing the parties to give up an exclusive sales territory.   [6] See 3.1–3.1.8 below.

[7] [1988] OJ L69/21, [1984] 4 CMLR 527, CMR 10,956, described at 3.1.4 below. See also *Mitchell Cotts/Sofiltra* [1987] OJ L41/31, [1988] 4 CMLR 111, CMR 10,852, analysed at 3.1.1 below; and *Boussois/Interpane* [1987] OJ L50/30, [1988] 4 CMLR 124, CMR 10,859, analysed at 3.1.3 below.

manufacture was essential to achieve various benefits: it provided an incentive to the licensee to invest and enabled it to use the new technique. Even so, the Commission found that the clause had the object of restricting competition in light of the special facts of the case, which it did not specify.[8] The Commission also granted an exemption in *Delta Chemie/ DDD*,[9] another agreement involving an exclusive manufacturing licence and some restriction of sales outside the territory, although the licensee could not have competed without a licence and there was considerable inter-brand competition. The finding that the exclusivity came within the prohibition of article 85(1) was not based on any economic analysis at all.[10] The Commission was, however, trying to establish the need for a group exemption for know-how licensing, so was predisposed to exempt such licences rather than clear them. Nevertheless, they remain the only Commission precedents on pure know-how licences, save for *Moosehead/ Whitbread*.[11] The patent licensing cases are even older. Fortunately, in *Nungesser*[12] and other cases, the Court of Justice of the European Communities (hereafter called the 'Community Court' or 'the Court') has taken into account under article 85(1) that protection may have to be given to licensees to induce investment.

The Commission has tried to overcome the problems of invalidity that arise from the wide scope it gives to article 85(1) by granting block exemptions.[13]

1.1.1 Drawbacks of the Commission taking a Broad View of Article 85(1) and Granting Exemptions under Article 85(3)—Invalidity of Contract and Comfort Letters

The Commission's unwillingness to clear agreements that, when applying article 85(3), it accepts increase competition has caused great difficulty for

---

[8] The special circumstances may have been the existence of other processes for preserving yeast when frozen. The Commission considered at para. 29 that these prevented the justification of the agreement as necessary to induce the dissemination of 'new technology', although it considered that an exclusive manufacturing licence confined to the United Kingdom, coupled with a licence to sell throughout the Common Market, was essential to induce Jus-Rol to take a licence and provide a useful product within the scope of art. 85(3). The existence of substitutes should have been a reason for clearing the agreement as not infringing art. 85(1), rather than the opposite.

[9] [1988] OJ L309/34, [1989] 4 CMLR 535, [1989] 1 CEC 2254, paras. 41–2, described at 3.1.5 below.

[10] Licensor and licensee were distributing pharmaceutical preparations and toiletries, although in different geographic markets. As a result of the distribution agreement, both were selling the licensed product, but DDD could not have done so without the distribution agreement, so the agreement was not horizontal. See 1.4.2 below.

[11] [1990] OJ L100/32, [1991] 4 CMLR 391, [1990] 1 CEC 2127, analysed at 3.1.7 below.

[12] *Nungesser (L. C.) AG and Kurt Eisele* v. *Commission* (258/78), [1982] ECR 2015, [1983] 1 CMLR 278, CMR 8805, analysed at 2.2–2.2.5 below.

[13] Those for technology transfer are described at 3.4–3.4.2 below.

firms, since the Commission has exclusive power to exempt, but lacks sufficient resources to grant many individual exemptions.[14] If an agreement be notified, the most probable result is that the Commission closes its file after sending a comfort letter. A comfort letter stating that the agreement is outside the prohibition of article 85(1) is helpful when enforcing the agreement in a national court because the court may take the comfort letter into account, although it is not bound by the Commission's view.[15] Where, however, it gives as its reason that the market share of licensor and licensees does not exceed 5 per cent, this may be unhelpful once the technology has been proved valuable and the parties' market shares have increased.

A letter stating that the agreement merits exemption used to make it harder to enforce the agreement, because it implies that the agreement does infringe article 85(1) or exemption would not be needed. A national court has no power to grant an exemption, so would probably adjourn the case for the Commission to decide whether to grant one.

Since *Automec II*,[16] however, it seems that the Commission may be under a duty to proceed to a formal decision under article 85(3) if so required, as that function is within its exclusive remit. In addition, just before he ceased to be the Member of the Commission in charge of competition policy, Sir Leon Brittan promised that the Commission would not depart from its comfort letters unless the parties had given false information or conditions had changed.[17] The problem is that an increase in market share may be perceived as a change in conditions.[18]

It may be argued, moreover, that to enforce an agreement that a court considers is likely to be exempted by the Commission does not amount to

---

[14] Individual exemptions can be granted only by a formal decision and the Commission has never been able to adopt many individual decisions each year. In the four-year period from 1990–3 the Commission managed an average of 4 individual exemptions per year, although in 1994 it achieved a record 17 individual exemptions, many of them concerned with joint ventures. The number was down again in 1995. Only a single technology licensing agreement has been the subject of a formal individual decision since the adoption of the know-how regulation in 1988. See *Moosehead/Whitbread*, n. 11 above, analysed at 3.1.7 below.

[15] *SA Lancôme and Cosparfrance Nederland BV* v. *Etos BV and Albert Heijn Supermarket* (99/77), [1980] ECR 2511, [1981] 2 CMLR 164, CMR 8714, para. 11.

[16] *Automec II—Automec Srl* v. *Commission* (T–24/90), [1992] ECR II–2223, [1992] 5 CMLR 431, para. 75.

[17] Speech on 'The Future of EEC Competition Policy' (Brussels: Centre of European Policy Studies, 7 December 1992). The relevant excerpt is reproduced in **Piet Jan Slot** and **Alison McDonnell**, *Procedure and Enforcement in EC and US Competition Law—Proceedings of the Leiden Europa Instituut Seminar on User-Friendly Competition Law* (London: Sweet & Maxwell, 1993), 120.

[18] See **Valentine Korah**, *Introductory Guide to EC Competition Law and Practice* (5th edn., London: Sweet & Maxwell, 1994), at 6.3.4.

granting an exemption,[19] but this has not been established. Sometimes, the Commission issues a neutral letter stating that it sees no reason to intervene under article 85 generally.

Agreements that have been notified with a request for exemption do not enjoy provisional validity.[20] Even if they were made before regulation 17 came into force and count as old, they lose their provisional validity once the Commission has sent a comfort letter stating that the file is being closed, since thereafter there is no likelihood of a retrospective exemption.[21] Consequently, the commercial risk for the licensor of investing in r & d and for the licensee in tooling up and making a market[22] may be aggravated by doubts about the validity of important provisions in the licence agreement.

I am optimistic that in the future more agreements may be cleared than hitherto.[23] Moreover, some files are disposed of by the Commission's staff sending a comfort letter saying that the agreement does not come within the prohibition of article 85(1).[24]

---

[19] Indeed, the Commission so submitted in *De Norre* v. *Brouwerij Concordia* (47/76), [1977] ECR 65, at 89, [1977] 1 CMLR 378, CMR 8386, although the Court avoided the question. See **René Joliet**, 'Trademark Licensing Agreements under the EEC Law of Competition' [1983–5] *Northwest J of Int'l. L & Bus.* 755, 764, n. 28.

[20] *Brasserie de Haecht SA* v. *Wilkin-Janssen (No 2)*, (48/72), [1973] ECR 77, [1973] CMLR 287, CMR 8170, paras. 25–7. According to the Court in *Société La Technique Minière*, n. 1 above, at 250 in the ECR, Community law requires the excision of the provisions that infringe art. 85, and whether what remains is enforceable is a matter for national law. The position under English law has been only partly worked out.

The judgment in *Brasserie de Haecht* was confirmed in *Delimitis* v. *Henniger Bräu AG*, n. 1 above, para. 49.

[21] *Lancôme*, n. 15 above, at para. 18. It is anomalous that when the parties have satisfied the Commission by altering their agreement sufficiently to persuade it to close its file, even an old agreement loses its provisional validity and any terms restrictive of conduct may become unenforceable.

Even at a technical level, the reasoning must be wrong. A letter that deprives an agreement of provisional validity affects the rights of the parties, so must be a decision, which should set out sufficient reasons as required by art. 190 and be subject to appeal: *Cimenteries—re Noordwijks Cement Accoord: Cimenteries CBR Cementbedrijven NV* v. *Commission* (8–11/66), [1967] ECR 75 at 91, [1967] CMLR 77, CMR 8052.

Moreover, since *Automec II*, discussed at text to n. 16 above, the Commission may have to reopen a file it has closed if the parties ask it to finalize its decision under art. 85(3). So the reasoning of the judgment in *Lancôme* may no longer be applicable.

[22] Some licensees also invest in developing the technology from a prototype for use on a commercial scale.

[23] See 1.4.2.1 and 3.1.8 below.

[24] Several officials confirm this.

Some practitioners still report that firms receive more comfort letters stating that the conditions of art. 85(3) are met and that the file is being closed, than that the agreement does not infringe art. 85(1). If a letter stating that the agreement merits exemption were produced in a national court deciding whether to enforce the agreement, the court might infer that the agreement was caught by art. 85(1), and that it could not enforce it until an actual exemption was adopted. It might have to adjourn for over a year and the defendant would then try to

The Commission is currently preparing a Green Paper expressing its views on vertical restrictions. The views of many officials are deeply held and diverse. Some think that vertical restrictions, especially exclusive territories and ancillary export bans, have been important in keeping the Common Market divided. Others think that the protection they may bring to firms deciding to incur sunk costs[25] are important in inducing innovation and its dissemination.

Since the know-how regulation granting a group exemption to a wider group of agreements than were covered by the patent regulation, there have been hardly any decisions on technology licences. The one decision is *Moosehead/Whitbread*,[26] which relates to what a lawyer might call a trade mark licence, and a businessman an industrial franchise.

1.1.2 Effects of Invalidity under English Law

In *Chemidus Wavin*,[27] the Court of Appeal had to decide whether a minimum royalty clause in a patent licence could be enforced when there were some provisions that also favoured the licensor but that might infringe article 85(1). Buckley LJ, with whom Orr LJ, concurred, said:

> 18. So, the position appears clearly to be this, that where in a contract there are certain clauses which are annulled by reason of their being in contravention of Article 85, paragraph (1), of the Treaty, one must look at the contract with those clauses struck out and see what the effect of that is in the light of the domestic law which governs the particular contract. In the present case, we have to consider what effect the invalidity, if any, of the clauses in the licence agreement by reason of Article 85 would have upon that contract as a whole. Whether it is right to regard the matter as one of severance of the contract or not, I do not think it is necessary for us to consider now. I doubt whether it is really a question of severance in the sense in which we in these courts are accustomed to use that term in considering whether covenants contained in contracts of employment and so forth are void as being in restraint of trade, and, if they are to any extent void, whether those covenants can be severed so as to save part of the covenant, although another part may be bad. *It seems to me that, in applying Article 85 to an*

---

persuade the Commission that an exemption would not be appropriate, although it might hold that enforcing the agreement does not amount to granting an exemption, as suggested in the text to n. 19 above.

[25] Costs are sunk when the investment has no other use, or to the extent that the next best use is less valuable. It may not be worthwhile to incur them unless the investor can appropriate the benefit—can reap where he has sown. Seen at the time the commitment to investment is made, protecting such an investor from competitors may be pro-competitive. See 1.3.1 below.

[26] Cited at n. 11 above and analysed at 3.1.7 below.

[27] *Chemidus Wavin* v. *TERI* [1978] 3 CMLR 514, [1977] FSR 181.

English contract, one may well have to consider whether, after the excisions required by the Article of the Treaty have been made from the contract, the contract could be said to fail for lack of consideration or on any other ground, or whether the contract would be so changed in its character as not to be the sort of contract that the parties intended to enter into at all. [my emphasis]

In *Chemidus Wavin*, the licensor was suing for a minimum royalty and the clauses that might be void were all in its favour. It made commercial sense to enforce what was left. The justice of the case might be different when some of the clauses that might be void favoured the licensee, which might not have agreed to pay the royalty or even to invest in the necessary equipment or marketing without, for instance, an exclusive licence. Its promise to pay royalties might have been in return in part for the technology licensed and in part for the protection that might be void by virtue of article 85(2). Without the promised protection it might have been willing to pay less or might, even, have declined to take the licence.

Were the licensor to sue in an English court for the royalties, the consideration would not wholly have failed, since the licensee did receive the benefit of the licence. The licensee would have to argue that the contract was so changed in its character that it was not one into which it had intended to enter.

In *Chemidus Wavin*, Goff LJ, as he then was, the co-author of a famous book, **Goff and Jones** on *The Law of Restitution*,[28] said:

> 30. . . . it seems to me that clause 4 [whereby the licensee agreed to pay a minimum royalty] is an independent promise supported by consideration and wholly unaffected by the invalidity, or possible invalidity, of other provisions in the agreement, and it seems to me to fall plainly within the principle laid down by Salter L.J. in *Putsman* v. *Taylor* [1927] 1 K.B. 639:
>
> > 'If a promisee claims the enforcement of a promise, and the promise is a valid promise and supported by consideration, the court will enforce the promise, notwithstanding the fact that the promisor has made other promises, supported by the same consideration, which are void and has included the valid and invalid promises in one document'.
>
> Here as it seems to me, the appellants made a valid promise to pay minimum royalties, supported by the licence which they obtained.

Goff LJ's quotation from *Putsman* v. *Taylor* may, however, be misleading, in relation to liability for royalties. Salter LJ's statement was made obiter, in that the covenant not to compete in *Putsman* v. *Taylor* that may have been void as being in restraint of trade was itself in issue, and no attempt was being made to sever the consideration as between that and the defendant's service as an employee.

[28] **Lord Goff of Chieveley** and **Gareth Jones**, *The Law of Restitution* (4th edn. ed. by Gareth Jones, London: Sweet & Maxwell, 1993).

Moreover, in *Chemidus Wavin*, the promises alleged to infringe article 85(1) were all for the benefit of the licensor, who was suing. In my hypothetical example of a licensee being sued for the royalties, the promise of an exclusive territory, which might infringe article 85(1), was for the benefit of the licensee being sued. The consideration would not wholly have failed: the licensee might be able to produce and sell the product, although the prices obtainable if other licensees might compete would be less and might not warrant his investment in tooling up and making a market. An exclusive licence might represent the second case in the passage I have italicized in Buckley LJ's judgment, where the contract without an exclusive territory was so changed in character that the defendant might not have entered into it. Nevertheless, this phrase was narrowly construed by Waller LJ in the passage I have italicized from his judgment in *Alec Lobb*,[29] described at the end of 1.1.2 below.

What is clear is that in *Chemidus Wavin* all the members of the court treated the precedents on severance in restraint-of-trade cases as relevant.

In a case involving only English law, *Alec Lobb*, the Court of Appeal, obiter but unanimously, supported the views of Goff LJ. An oil company bought and leased back a petrol station in return not merely for rent, but also for an obligation to buy petrol solely from the landlord. The second obligation was arguably void as being in restraint of trade. On the facts, none of the judges considered that the *solus* arrangement in issue was in unreasonable restraint of trade.

Nevertheless, the issue of severance had been fully argued, so they each gave his view. Their views may not bind lower courts absolutely, but are of strong persuasive authority. Dillon LJ said[30] that the tie was neither the sole consideration nor the sole object of the transaction. Although he considered that Total would not have granted a lease-back without the *solus* arrangement, that was not enough to prevent severance.

Dunn LJ thought that the agreement was severable provided that the valid promises were supported by some consideration, even though the consideration also supported the invalid promise, unless the invalid promise was substantially the whole or main consideration.

Waller LJ looked to the part of Buckley LJ's judgment in *Chemidus Wavin* that I have italicized, but concluded[31] that:

> Even though Total would not have entered into the contract without the tie it remained a contract for letting a petrol station and it was at a rent which was not nominal. *It was therefore the sort of contract which the parties intended to enter into.* [my emphasis]

---

[29] *Alec Lobb Ltd* v. *Total Oil GB Ltd* [1985] 1 All ER 303, [1985] 1 WLR 173.
[30] *Ibid.* 311E and F (All ER).   [31] *Ibid.* 320 (All ER).

This interpretation reduces Buckley LJ's qualification almost to nothing. Contracts may be classified in many ways. The description of the agreement as a lease and lease-back is only one way: commercial men might classify it as a *solus* agreement to finance the premises. Why should the legal classification determine the outcome?

One day, it may be argued that the Court of Appeal was wrong to treat invalidity under article 85(2) in the same way as contracts in restraint of trade. The latter are not illegal, but merely unenforceable. Infringements of article 85(1), however, are prohibited by article 85(1) as a matter of public policy. The sanction is not intended to ensure fairness as between the parties, but to discourage anti-competitive agreements that may harm the economy as a whole. Often the licensee will have enjoyed the benefit of an exclusive territory *de facto*.

It is not the licensee's interest that needs protection by sanctions, but that of purchasers of the product made with the help of the technology. It may be argued that they have had to pay more than they would have done in the absence of an exclusive licence. Inability to obtain any, or full, royalties might discourage innovators from granting more protection than they think the Commission or a national court would accept and it might be too risky to exploit some inventions without enough protection. Uncertainty as to what is enough may unduly reduce the incentives to take a licence or to conduct the original research.

I will suggest at 1.4.1 and 2.2 *et seq.*, however, that exclusive territories and the associated export bans rarely restrict competition, unless the licensee could have competed with the licensor without a licence. A licensor usually benefits if licensees operate on smaller margins. If they are allowed to compete with each other, prices to their customers may be reduced and more of the product sold. So the licensor has an incentive to offer as little protection as possible between the licensees in his network. It is commercially sensible for it to protect one licensee from the others only if it considers that this is necessary to induce investment in tooling up and developing the market.

The licensor is nearer the market than is a national judge or an official in a competition authority, and is backing its judgement with its likelihood of profits. It is more likely to be right. Where the parties to a license could not have competed without a licence—when the agreement is vertical—there is no need for competition law to intervene.

If there is any doubt about the legality under Community law of clauses in favour of a licensee when drawing up a contract, thought should be given to splitting the consideration—so much for the exclusive territory and associated export bans and so much for the know-how.

1.2 THE EC COMMISSION'S ATTITUDE TOWARDS EXCLUSIVE TECHNOLOGY LICENCES IS BECOMING INCREASINGLY POSITIVE

Since the Commission started to prepare the know-how regulation, it has recognized that competition stimulates innovation and innovation stimulates competition. UNICE and others stressed the need for legal certainty and, in the first recital to the know-how regulation, the Commission stated that it also accepts the need for legal certainty. Recital 12 to the technology transfer regulation refers to making licensors more willing to grant licences and licensees to undertake investment, which may well be partly because of the certainty of enforcing exclusive provisions. When the decision to invest in innovation is made by the licensor, it must expect to be able to appropriate to itself the value of benefits it generates; while, at the time when the contract is being negotiated, the licensee must also expect to be able to appropriate the benefits of its investments in perfecting the process, tooling up, and developing a market. Otherwise, the one will not innovate and the other will not commit itself to paying royalties, tooling up, and developing a market.

In *Nungesser*,[32] the Court went a long way under article 85(1) in recognizing the need to induce investment by licensor and licensee by granting open exclusive licences.

In 1984, the Commission granted a group exemption[33] for pure patent licences and for licences including both patents and know-how that was secret and permitted a better exploitation of the licensed patents, provided that the licensed patents were necessary for achieving the objects of the licensed technology.

It was objected by industry that although such a group exemption was useful, it was very limited and should be extended to include pure and other mixed know-how licences. The position was made more acute in *Boussois/Interpane*[34] when the Commission held that the patent regulation did not apply for two reasons: first, because in that case the know-how licensed dominated the technology and did not merely permit a better exploitation of the licensed patents[35] and, secondly, because there were no

---

[32] Cited n. 12 above, and analysed at 2.2–2.2.5 below, paras. 56 and 57. See the Commission's narrow interpretation of the precedent in recital 10 to the Technology Transfer Regulation.

[33] Reg. 2349/84, [1984] OJ L219/15, analysed by **Valentine Korah**, *Patent Licensing and EEC Competition Rules—Regulation 2349/84* (Oxford: ESC Publishing, 1985).

[34] Cited n. 7 above, and analysed at 3.1.3 below.

[35] In para. 19 of the decision, the Commission observed that there were several Member States where there was no patent, and that the licensee was not bound to exploit the patents or pay royalties throughout the term of the contract. It continued:

'20. The agreement is therefore not covered by Regulation 2349/84, and until such

patents in some Member States for which territorial protection was provided. A licence to manufacture in facilities provided in France by the licensor was treated by the parties as a restriction on manufacturing elsewhere and, in so far as this related to Member States where there were no patents, was blacklisted by article 3(3) and (10).[36]

There were fears, first, that for the agreement to qualify the know-how had to be ancillary to the patent licensed and not the other way round, although the word 'ancillary' did not appear in the final draft of the patent regulation. The test was not workable and the Commission seems to have been unhappy with it. When both patent and know-how are necessary, they are complementary and neither is ancillary to the other.

On the second view, since it is unusual for those paying for innovation to take out patents in all Member States,[37] many licences failed to come within regulation 2349/84, as there were usually restrictions on making or selling outside the territory in countries where there was no patent protection.[38] Few notified licences were able to benefit from the group exemption.

> time as there is a block exemption regulation specifically for pure know-how agreements or mixed agreements in which the know-how component consists of a body of knowledge that is crucial for the exploitation of the licensed technology and not just a factor permitting a better exploitation of the patents, the restrictions of competition involved in the agreement require individual exemption.'

[36] The Commission accepted the views of the parties and treated the licence to produce only in the factory provided in France by Interpane as a restriction on manufacturing or selling outside France. See 3.1.3, n. 8 below. Such restrictions would not be exempted under art. 1(1) of the patent licensing reg., even if the know-how had been ancillary to the patents, in so far as there were no patents in some of those countries. Informally, officials explained the case on the basis that the opposition procedure cannot apply if there are blacklisted clauses, but the published decision can be understood to mean that, where the agreement does not qualify as a patent licence, the opposition procedure in the reg. exempting patent licences cannot be used to extend its scope.

[37] There are many reasons for this. First, some Member States of the Communities have only recently joined the European Patent Convention. So, many patents still in force elsewhere were not applied for in those countries. Now all the Member States of the EC have joined the Convention. Secondly, even if one applies under the European Convention, the fees are higher if patents are wanted for more countries and the patent applied for becomes operative only after expensive translations have been made into the local language. Thirdly, it is easier to obtain patents in some countries than others. Fourthly, some products are useful only in certain areas. Dripper pipes, for instance are used only where the climate is hot and evaporation is a serious problem. So it would not be worth obtaining a patent in Scandinavia. On the other hand, all three group exemptions for technology transfer have applied not only to patents, but also to applications and even to applications taken out within a year after the licence, at least if the application is granted.

[38] Jean-François Verstrynge, then the member of Mr Sutherland's cabinet responsible for arts. 85 and 86, confirmed that several notified agreements failed to enjoy the benefit of the group exemption for this reason: **Jean-François Verstrynge**, in **Barry Hawk** (ed.), *Thirteenth Annual Proceedings of the Fordham Corporate Law Institute* (New York: 1986), 311, 314. Until *Boussois*, this was not widely realized by practitioners.

1.3 THE DANGER OF RELYING ON OLD CASES

1.3.1 The Commission is More Willing to Analyse *Ex Ante*

In the last decade, competition policy has become less interventionist than it was in the 1970s and early 1980s.[39] In old cases like *AOIP* v. *Beyrard*[40], the Commission was concerned, *inter alia*, about package licences, where the patentee could add new patents and thereby extend the period during which the restrictions continued to operate and for which royalties were due. Such an agreement may be perceived as a tie of the period when the patentee may have had market power because of the exclusive basic intellectual property right, to the period when it would enjoy less protection under an improvement patent.[41] Its obligations and the restrictions accepted would make it difficult for the licensee to compete when the original patent expired. The Commission saw its role as including the protection of licensees from the consequences of making a bad bargain.

By 1987, the Commission took the view that the parties should be able to make their own decisions about royalties.[42] The licensor may want to share the risk of premature disclosure with the licensee. Presumably, if the licensee is going to pay for longer than the know-how remains secret or the patent remains valid, it will be prepared to pay less in the early years or demand some other benefit. It may be easier for a small licensee to raise credit if it incurs little liability for royalties on the products made at the beginning, when its investment is at greatest risk. This view has been maintained in recital 21 and article 2(1)(7) of the technology transfer regulation.

The Commission used to consider that most, although not all, patent licences were horizontal once both licensor and licensee are manufactur-

---

[39] See **Dr. Hartmut Johannes**, 'Technology Transfer under EEC Law—Europe between the Divergent Opinions of the Past and the New Administration: A Comparative Law Approach', in **Hawk** (ed.), *Ninth Annual Proceedings of the Fordham Corporate Law Institute* (New York: 1982), 65, for the policy of the 1970s. See, shortly, 1.1 above.

[40] [1976] OJ L6/8, [1976] 1 CMLR D14, CMR 9801, criticized by **Valentine Korah**, 'Patents and Competition Law: Recent Decisions of the European Commission' [1976] 1 *EL Rev.* 185.

[41] See the US case of *Brulotte* v. *Thys*, 379 US 29 (1964). The US agencies would probably not have intervened in such a situation since 1981.

[42] See the Commission's submissions to the Court in *Ottung* v. *Klee & Weilbach A/S and Thomas Schmidt A/S* (320/87), [1989] ECR 1177, [1990] 4 CMLR 915, [1990] 2 CEC 674, at 1182–3 of the ECR, quoted at 7.3.7 below together with art. 2(1)(7) of the technology transfer reg., analysed in the light of recital 21, which whitelists obligations to pay royalties 'for the periods and according to the methods freely determined by the parties'. The Court in *Ottung* was not quite as liberal as the Commission's submission, although it accepted that royalties might outlast the life of the patent.

ing.[43] Now it is less sure. Sometimes, it analyses a situation *ex ante*, when the agreement is negotiated and commitments made to invest in tooling up, paying royalties, and making a market, rather than *ex post*, when it assesses its effect on competition.[44] In this, the Court has led the way in a series of cases[45] in which it has held that various ancillary restrictions did not infringe article 85(1), if without them the parties would not have committed themselves to making risky investment.

1.3.2 Know-how is More Vulnerable and Needs More Contractual Protection than Patented Technology

Moreover, the Commission's formalistic views about know-how have changed dramatically. It used to believe that, since know-how was not

---

[43] See **Johannes**, n. 38 above, 65, 83. See also **James Venit**, 'In the Wake of *Windsurfing*: Patent Licensing in the Common Market', in **Hawk** (ed.), *Thirteenth Annual Proceedings of the Fordham Corporate Law Institute* (New York: 1986), 517.

[44] For examples from the technology transfer reg. see: **(1)** recital 12, which states that territorial protection makes licensors more willing to grant and licensees to receive licences; **(2)** the definition of 'substantial' in art. 10(3), analysed at 6.1.1.2 below; **(3)** the clearance of the post-term use ban, analysed at 7.3.3 below; **(4)** whether quality specifications or a tie are required to be assessed at the time when the contract is signed, art. 4(2)(a) analysed at 7.3.5.2 below; **(5)** restrictions on challenging the validity of patents are now subject to the opposition procedure, whereas they used to be blacklisted: see 7.3.6.2 below; **(6)** the Commission's view that royalties are a question entirely for the parties, expressed in recital 21 and discussed at 7.3.7 below; **(7)** art. 2(1)(10) permits a most-favoured-licensee provision, discussed at 7.3.10 below; **(8)** art. 3(4) blacklists customer restrictions only when the parties were competing manufacturers at the time of the contract; **(9)** art. 7(4) enables the Commission to withdraw the exemption in certain circumstances when the parties were competing manufacturers at the time of the grant of the licence: see 9.1.3.4 below; **(10)** the Commission's perception of the need for territorial protection when licensor and licensee are making commitments to invest, expressed in recital 12: see 4.1.3 below.

Examples taken from Commission decisions include *Mitchell Cotts/Sofiltra*, n. 7 above, paras. 18–24. Contrast some of the Commission's decisions adopted as recently as late 1988, such as *Delta Chemie*, n. 9 above, discussed at 3.1.5 below.

In many of its decisions, however, the Commission has perceived the situation *ex post* and treated as potential competitors those who could not have competed without a licence: e.g., *BBC Brown Boveri—Re the Agreements between BBC Brown Boveri and NGK Insulators Ltd* [1988] OJ L301/68, [1989] 4 CMLR 610, CMR 11,305, para 19.

In *Optical Fibres*, [1986] OJ L236/30, CMR 10,813, para. 46, however, the Commission stated that either of the joint ventures to which Corning granted a licence would not have been caught by art. 85(1) because they had complementary technologies: it was only the existence of several joint ventures with a common supplier of technology that led to the exclusive licences requiring exemption. See also *Odin* [1990] OJ L209/15, [1991] 4 CMLR 832, [1990] 2 CEC 2066, where a licence to a joint venture that was exclusive for its field of use was cleared.

The Court has more often considered the need for incentives at the time commitment is made to investment. See the doctrine of ancillary restrictions explained at 1.4.1 below, the later cases on free movement, explained at 2.1.3, and the judgments on exclusive licences at 2.2.1 below.

[45] For the greater willingness of the Court to recognize the need for incentives to investment, see 1.4.1, 1.4.2, and *Nungesser*, n. 12 above, discussed at 2.2–2.2.5 below.

protected by an exclusive right, the contract deprived the licensee of his freedom to do what would otherwise be legal, and should be looked on with suspicion.[46] By the time the know-how regulation was drafted, it accepted that know-how licences were important and that the very vulnerability of know-how made it necessary for the licensor to impose obligations by contract if it was not to lose the benefit of its innovations.[47] The change of view is important as pure patent licences are rare, save where granted to resolve a dispute.

1.4 SHOULD FEWER AGREEMENTS BE FOUND TO INFRINGE ARTICLE 85(1)?

1.4.1 The Distinction between Naked and Ancillary Restraints

In the United States, the courts early developed a distinction between ancillary restrictions necessary to make viable some pro-competitive transaction which did not necessarily infringe the antitrust rules and naked restraints of trade that had no purpose but to restrict competition, the reasonableness of which would not be considered by the courts. In *Addyston Pipe & Steel Co.* Judge Taft said that, for instance, covenants not to compete with a business sold as a going concern were necessary if businesses were to be sold with their goodwill,[48] and the reasonableness of their scope in time and space could be judged by reference to their main object. When clothed by a lawful transaction, ancillary restraints were valid, provided they were no more restrictive than was reasonably necessary to make the main transaction viable. On the other hand, to decide whether the prices fixed by a cartel were no higher than was reasonable would exceed the judicial function—would be 'to set sail on a sea of doubt'.[49] Such naked restraints of competition are almost certain to be anti-competitive, and naked price fixing, market allocation, and

---

[46] In *Reuter/BASF* [1976] OJ L254/40, [1976] 2 CMLR D44, CMR 9862, the Commission seemed to deny that one could sell know-how. An inventor should be free to continue innovating and exploiting his inventions at least from a reasonable time after transferring his know-how.

Restrictions on using the technology after the licence expires were looked at as restrictive, which seems to imply that one might not lease know-how. This view is expressly rejected by recital 20 and art. 2(1)(3) and 2(1)(4) of the Technology Transfer Reg.

See generally, **Venit**, n. 43 above, and **Johannes**, n. 39 above, at 88.

[47] See the recitals to the technology transfer reg., considered at 4.1–4.1.4 below.

[48] He gave 4 examples: (1) a covenant by a partner or (2) by a retiring partner not to compete with the partnership; (3) one by a servant not to compete with his master's business; and (4) a covenant by the vendor of property or a business not to compete with the buyer in such a way as to derogate from the value of the property or business sold.

[49] *United States* v. *Addyston Pipe & Steel Co.*, 85 Fed. 271, 284 (6th Cir. 1898), at 284, *affirmed* 175 US 211 (1899). Judge Taft later became President of the USA.

collective boycotts can be condemned with little market analysis, without risking much loss of efficiency.

The Community Court has adopted a very similar approach to ancillary restraints. The only difference seems to be that it does not enquire whether the transaction made viable by the ancillary restraint is, on balance, pro-competitive.[50] As early as *Société La Technique Minière*[51] the Community Court ruled that if a French manufacturer of earth-moving equipment could not penetrate the German market without an exclusive distributor there, the exclusivity would not infringe article 85(1). *Remia*[52] was very similar to one of the examples given by Judge Taft and, indeed, to the English cases on restraint of trade at common law. The Court ruled that a covenant by a seller not to compete with a business sold with its goodwill does not infringe article 85(1) as long as it is reasonably limited in time and space. It implied at paragraph 34 that the Commission has a wide discretion in assessing the reasonableness of the restrictions under article 85(1).

There have been several judgments on the licensing and exercise of intellectual property rights where investments often have to be made in advance and are viable only if the investor is sufficiently protected from the competition it makes possible to be able to appropriate the benefit to itself. In *Nungesser*,[53] the Court quashed the Commission's finding that an exclusive licence of plant breeders' rights infringed article 85(1) because the Commission had not considered whether 'open exclusivity' might be justifiable on the ground that investment by both parties was necessary to develop the technology and a market in Germany. The Court rejected the Commission's practice of treating restrictions of conduct that are commer-cially important as necessarily restricting competition. Such restraints may be necessary to induce investment. This should have been investigated by the Commission.

In *Coditel II*[54] the Court went further in relation to performing rights. Advocate General Reischl had stressed that, since the doctrine of exhaustion[55] does not apply to the free movement of services, the holder of

---

[50] The US courts, however, in fact do far less balancing than they are widely thought to do. See **Barry Hawk**, in **Hawk** (ed.), *Fourteenth Annual Proceedings of the Fordham Corporate Law Institute* (New York: 1987), 738. [51] N. 1 above, at 250 (ECR).

[52] *Remia BV and Others* v. *Commission* (42/84), [1985] ECR 2545, [1987] 1 CMLR 1, CMR 14,217. [53] N. 12 above, at paras. 56 and 57. See 2.2–2.2.5 below.

[54] *Coditel SA and Others* v. *Ciné Vog Films SA and Others (No 2)* (262/81), [1982] ECR 3381, [1983] 1 CMLR 49, CMR 8862.

[55] In many cases, such as *Centrafarm BV* v. *Sterling Drug Inc.* (15/74), [1974] ECR 1147, [1974] 2 CMLR 480, CMR 8246, discussed at 2.1.2–2.1.3 below, the Court held that, once goods were sold in one Member State by or with the consent of the holder of patents, trade marks, or copyright, etc., its rights were exhausted by the rules for the free movement of goods and could not be exercised to restrain the import or sale of goods protected by the industrial or commercial property right in another Member State.

the performing rights in a film, or its exclusive licensee, may exercise its copyright in one Member State to exclude the transmission of performances from another. Therefore, for the Court to hold that the original agreement granting exclusive territories does not *in itself* restrict competition would confer absolute territorial protection. Despite the Court's hostility to absolute territorial protection, he suggested that the Court should rule that the agreement would not be incompatible with article 85(1) if, without an exclusive territory, no licensee could have been found for the territory in question. The Court followed his suggestion and ruled that:

> 11. . . . the film belongs to the category of literary and artistic works made available to the public by performances which may be infinitely repeated and the commercial exploitation of which comes under the movement of services, no matter whether the means whereby it is shown to the public be the cinema or television.
>
> . . .
>
> 16. The characteristics of the cinematographic industry and of its markets in the Community, especially those relating to dubbing and subtitling for the benefit of different language groups, to the possibilities of television broadcasts, and to the system of financing cinematographic production in Europe serve to show that an exclusive exhibition licence is not, in itself, such as to prevent, restrict or distort competition.

Unless the right to restrain performances were recognized the holder or licensee would not be able to obtain a fair reward for creating a film. Consequently the right is part of the specific subject matter of the intellectual property right, and an exclusive licence infringes article 85(1) only if excessive charges are made to exhibitors.[56]

In *Erauw-Jacquéry*,[57] the Court went very far in relation to plant breeders' rights over basic seeds supplied for propagation to growers of the certified seed. It should be remembered that, for the breeders' right to

---

[56] At para. 19, the Court tried to help the referring court to consider how far exclusivity was required by giving a series of criteria that are impossible to apply, but relate to whether, after the event, profits have turned out to be too high. Not only does this reduce the incentive to investment, which must be encouraged *ex ante*, as the Court recognized at para. 16, quoted in the text just above this note, it is impossible to define appropriate criteria to decide what reward is appropriate. Most films make little profit, but the production industry continues because an occasional film makes huge profits. Since producers are unable to tell in advance which films will make profits, they make many, and the costs of making the ones that attract little demand must be covered by the profits on the successful films. One film is a by-product of the others.

[57] *Erauw-Jacquéry (Louis) Sprl* v. *La Hesbignonne Société Coopérative* (27/87), [1988] ECR 1919, [1988] 4 CMLR 576, [1989] 2 CEC 637. Minimum prices imposed for seed of all species and not confined to those grown by selected propagators might, however, infringe art. 85(1) if they have the effects on competition and trade between Member States prohibited by the provision.

remain valid, the variety must remain distinct, uniform, stable, and useful. The Commission argued that the existence of the right includes the holder's right to control those who propagate the seed. The Court did not refer to the specific subject matter of the right but, at paragraph 10, did refer to *Nungesser*, to the investments required to develop basic lines of seed, and to the holder's need to obtain protection against improper handling of the seed:

> 10. . . . the development of the basic lines may involve considerable financial commitment. Consequently, a person who has made considerable efforts to develop varieties of basic seed which may be the subject-matter of plant breeders' rights must be allowed to protect himself against any improper handling of those varieties of seeds. To that end, the breeder must be entitled to restrict propagation to the growers which he has selected as licensees. To that extent, the provision prohibiting the licensee from selling and exporting basic seed falls outside the prohibition contained in Article 85(1).

Consequently, it ruled that at that stage of production, customer restrictions and export bans were justified and not contrary to article 85(1).[58] The Commission stated in its *Eighteenth Report on Competition Policy*[59] that, although the export ban was not directly in issue, the Court had made it clear that in the context of licences to a propagator of basic seed, such a clause did not infringe Article 85(1).

Two years later, in *Association of Plant Breeders of the EEC* (*Comasso*),[60] the Commission announced its intention to take a favourable decision on rather similar standard licensing agreements which included export bans on the propagator despite the horizontal nature of the arrangement. The Commission, however, considers that the precedent is confined to basic seed.[61]

---

[58] The precedent is interesting in relation to software licensing, which is not covered by the know-how or technology transfer reg., save where the software is ancillary to other know-how. See 5.1.4 below. There is no case law on software licences by the Court or the Commission, although a dir. has been adopted by the Council on the legal protection of computer programmes, [1991] OJ L122/42, which gives some guidance on the Commission's views.

[59] (Brussels: EC Commission, 1989), para. 103.

[60] Art. 19(3) Notice [1990] OJ C6/3, [1990] 4 CMLR 259.

[61] The judgment in *Erauw Jacquéry*, n. 57 above, may be extended to justify customer restrictions when the products are technical and need careful handling, for instance, installation. Nevertheless, it should be stressed that the Court referred only to basic seed before it has been propagated for sale as certified seed to farmers, although the reasoning was general, based on the need to invest in developing a line of seed. Basic seed deteriorates fast and can be grown only for one or two generations, depending on the variety, before ceasing to qualify as distinct, uniform, and stable, as required by the international convention under which plant breeders' rights are protected by national laws.

In *Volvo*[62] and *Renault*,[63] the Court again looked to the cost of designing both a car as a whole, and body parts in particular, and ruled that the exercise of design rights does not, *in itself*, restrict competition. Advocate General Mischo, in the *Renault* case, pointed out at para. 31 that the producer of the vehicle does not infringe article 86 by obtaining part of its reward on the sale of the car and part on the sale of the replacement part.

In these cases, the Court has stressed the need for a fair reward and for inducements to investment by the licensor in developing the technology and by the licensee in preparing to produce and developing a market to exploit it. It has not had to consider how much protection is required to enable the investor to appropriate enough of the benefit of its investment to make it commercially viable. It is not a finder of fact. Its power under article 173 of the EC Treaty to review the Commission's decisions is limited and, when issuing a preliminary ruling under article 177 of the Treaty, it does not apply the law to the facts. That is the task of the national court or tribunal referring a question of the construction of Community law to the Court.

Magill[64] involved copyright in a list of television programmes, a law that has been subject to much criticism on the ground that compiling the list requires neither originality nor effort and no incentive is required for publication of future programmes. The Court held that the television companies that held such copyrights were each dominant over the information reproduced in the lists, and it required a compulsory licence to be granted to a firm that wanted to publish a collective guide to all the programmes that could be received in Ireland and Northern Ireland. It is thought that the case was very exceptional and that compulsory licences will seldom be required. At 2.3, I will argue that if the holder of an important pharmaceutical patent were required to license the holder of an improvement patent, the incentive to the original innovation would be reduced.

There is a substantial literature advocating a realistic market analysis under EC law, partly on the basis of the Court's precedents and partly because of the need to be able to enforce ancillary restrictions if firms are to make investments.[65]

---

[62] *Volvo* v. *Erik Veng (UK) Ltd* (238/87), [1988] ECR 6211, [1989] 4 CMLR 122, CMR 14,498. See also **Valentine Korah**, 'Comment, No Duty to License Independent Repairers to Make Spare Parts: The *Renault, Volvo* and *Bayer & Hennecke* Cases' (1988) 12 *EIPR* 381.

[63] *Consorzio Italiano della Componentistica di Ricambio per Autoveicoli and Maxicar* v. *Régie Nationale des Usines Renault* (53/87), [1988] ECR 6039, [1990] 4 CMLR 265, [1990] 1 CEC 59. See also **Korah**, n. 62 above.

[64] *Magill—Radio Telefis Eireann (RTE) and Independent Television Publications Ltd (ITP)* v. *Commission* (C–241 & 242/91P), [1995] ECR I–743, [1995] 4 CMLR 718, [1995] 1 CEC 400. See 2.3 below.

[65] The following is a selection of writings on the subject: **René Joliet**, *The Rule of Reason in Antitrust Law* (The Hague: Martinus Nijhoff, 1967); **Joliet**, n. 19 above, 773; **Christopher**

1.4.2 The Distinction between Horizontal and Vertical Restraints

In the United States, the distinction between horizontal and vertical agreements was considered important by the middle of this century. Horizontal relationships are those between undertakings operating at the same level of the same trade or industry—in other words between actual or potential competitors. Vertical agreements are those made between firms at different stages of bringing a product to market: a manufacturer and its wholesaler or retail dealer; the holder of technology and the firm it licenses; the supplier of raw material or a component and the maker of the finished article.

Agreements between competitors may well be intended to raise prices and provide benefits only to the parties at the expense of those with whom they deal and of others who might have dealt had prices been at a more competitive level. They include naked[66] price-fixing and collective boycotts which are usually illegal in the United States.

In that country, only if the restriction is ancillary to some pro-competitive transaction and reasonably necessary to make it viable are horizontal agreements not treated as automatically contrary to section 1 of the Sherman Act. Some joint ventures entered into between competitors or potential competitors are valid: where, for instance, the risk is too great for a single firm to bear or the minimum efficient scale is large in relation to the expected demand, it is possible that not more than one of the parties would alone undertake the activity of the joint venture. In that situation, the parties may not be in a horizontal relationship and the venture may be treated as not anti-competitive. The Antitrust Division of the Department of Justice during the Reagan administration stated its intention not to

---

Bright, 'Deregulation of EC Competition Policy: Rethinking Art. 85(1)', in **Hawk** (ed.), *Twenty-Second Annual Proceedings of the Fordham Corporate Law Institute* (New York: 1995), 505; **Jonathan Faull**, 'Joint Ventures under the EEC Competition Rules' (1984) 5 *EL Rev.* 358, 362; **Ian Forrester** and **Christopher Norall**, 'The Laïcization of Community Law— Self Help and the Rule of Reason: How Competition Law is and could be Applied' (1984) 21 *CML Rev.* 11, and in **Hawk** (ed.), *Tenth Annual Proceedings of the Fordham Corporate Law Institute* (New York: 1983), 305; **Luc Gyselen**, 'Vertical Restraints in the Distribution Process: Strength and Weakness of the Free Rider Rationale under EEC Competition Law' (1984) 21 *CML Rev.* 647; **Stephen Kon**, 'Article 85, Para.3: A Case for Application by National Courts' (1982) 19 *CML Rev.* 541; **Valentine Korah**, 'The Rise and Fall of Provisional Validity The Need for a Rule of Reason in EEC Antitrust' [1981] 3 *Northwestern J Int'l. L & Bus.* 320, 340 ff.; **Valentine Korah**, 'EEC Competition Policy: Legal Form or Economic Efficiency' (1986) 39 *Current Legal Problems* 85; **Korah**, n. 62 above; **Mark Schechter**, 'The Rule of Reason in European Competition Law' [1982/2] *LIEI* 1.

A narrower application of art. 85(1) is now widely advocated. See the paper of the CBI, 'Refocusing the Competition Rules—Making the Single Market Work'.

[66] See Judge Taft in *Addyston*, n. 49 above. He contrasted ancillary restraints, that made possible some legitimate transactions and were 'clothed by a lawful arrangement', from 'naked restraints', which were not.

challenge some very large joint ventures on the ground that not more than one of the venturers would have carried on alone the task of the joint venture.[67]

An agreement between a brand owner or holder of technology, H, and its distributor or licensee, however, is usually vertical[68] and, if H decides that it must induce investment by the latter by protecting each dealer or licensee from the others, it is likely that the protection creates efficiencies. Other things being equal, H would sell more or would earn higher royalties if prices were forced down by competition. Even if H has market power, it has no reason to share any excess profits with its dealers or licensees. The lower their margin, the more H will earn. If, then, H protects them from each other, the most likely reason is that H wants to induce investment that pays the network as a whole, but would not be worth while for any one of the firms downstream unless it could appropriate the benefit of the investment to itself.[69]

If one dealer or licensee, A, promotes the brand as a whole by providing pre-sales services or perfecting the licensed process, another dealer or licensee, B, may take a free ride on A's investment to undercut him. If this is possible, A will be discouraged from making the investment in the first place as it will not be profitable to A individually. The brand owner or technology holder may, therefore, want to protect A, whether by giving him an exclusive territory, supported by export bans, or through imposing minimum prices on the other licensees or dealers, or otherwise.

---

[67] The business review letter about *Aero Engines* (1983) 45 *ATRR* 726 is an extreme example. Half the possible manufacturers in the world agreed to collaborate to design an engine for a particular size of aeroplane. Once one of them incurred the design costs, it would not be profitable for any of the others to enter the market as the demand would not support a further model and, consequently, the Department of Justice indicated that it would not take action under the antitrust laws. Any one of the joint venturers could enter the market, but not more than one.

Joseph Griffin, of the US law firm Morgan, Lewis, and Bockius, tells me that the attitude of the Department of Justice to joint ventures has not changed much under the Clinton administration. One of the factors that continues to be listed in various speeches is whether fewer than all the parties could have engaged in the venture. The Department of Justice rather quickly changes its focus to the issue of ancillary restraints. See the speech of J. Robert Kramer II, Chief, Litigation II Section, Antitrust Division, 'Contractual Joint Ventures: The Enforcement View' (7 Aug. 1995.)

For the distinction between horizontal and vertical restraints in relation to licensing, see the joint Department of Justice and Federal Trade Commission *Guidelines on Intellectual Property Licensing* (6 Apr. 1995). See 1.4.3 below.

[68] It would have a horizontal element if licensor and licensee were competitors or potential competitors without a licence. Cross licensing between the members of a technology pool might be horizontal, but might be justified as saving on the need to check intellectual property claims and to avoid blocking patents generally, whether or not third parties were able to join the pool. See 5.1.1–5.1.1.3.

[69] **Lawrence J. White**, 'Vertical Restraints in Antitrust Law: A Coherent Model' [1981] XXVI *Antitrust Bull.* 327.

Suppliers or licensors who give dealers or licensees more protection than needed will earn lower profits and be able to expand less than those who get the decision right. They will be punished by the market. So there is no need for authorities to intervene.[70]

Some economists in the United States have advocated *per se* legality for vertical restrictions.[71] Traditional economists have tended to condemn some vertical practices. A major drawback of interfering with vertical relationships is that it may distort markets. Control may be avoided by one firm in the vertical chain starting to perform the functions that might more efficiently be performed by others or by acquiring firms up- or downstream. Vertical integration may be desirable if it is more efficient to bring both activities under single control than to negotiate transactions through the market. Market pressures encourage firms to integrate only when this is more efficient than buying or selling. So there is no need for competition authorities to encourage vertical integration. Grundig acquired Consten after the Commission interfered with its contract with Consten,[72] its exclusive distributor for France. It is unlikely that this made the French market more competitive.

In judging the validity of vertical restraints, other lawyers and economists[73] would want to consider whether the market is one where the

---

[70] **Warwick A. Rothnie** and I analysed the economic arguments relating to vertical agreements more fully in *Exclusive Distribution and the EEC Competition Rules: Regulations 1983/83* (2nd edn., London: Sweet & Maxwell, 1992), ch. 1.

[71] See, e.g., **Lester G. Telser**, 'Why Should Manufacturers want Fair Trade?' (1960) 3 *J Law & Econ.* 86; **Robert Bork**, *The Antitrust Paradox* (2nd edn., New York: Basic Books, 1978), 288–98 and the app. to chs. 13 and 14; **Richard Posner**, 'The Next Step in the Antitrust Enforcement Treatment of Restricted Distribution: *Per se* Legality' (1981) 48 *U Chicago L Rev.* 6, 22–6, especially 25.

Frank Easterbrook considers that practices should be condemned by courts only if it is established that the defendant enjoys market power and that the practices may enrich it at the cost of consumers. In other situations, there is no need for the law to intervene. The market is a sufficient safeguard. If firms do have the ability and incentive to behave in an anti-competitive way, then the court should see whether competitors use different methods of production and distribution, in which case, again, there will be sufficient competition. If not, then the court should ask whether the evidence is consistent with a reduction in output. Moreover, if the plaintiff is a competitor of the defendant, there is some evidence that the practice is beneficial to consumers: **Frank Easterbrook**, 'Vertical Arrangements and Rule of Reason' (1984) 53 *Antitrust LJ* 135, and **Frank Easterbrook**, 'The Limits of Antitrust' (1984) 63 *Texas L Rev.* 1.

[72] See *Consten/Grundig* [1964] JO 2545, [1964] CMLR 489, considered at 2.1.1 below.

[73] George Hay wrote a good, balanced, and short summary of both sides of the argument: **George Hay**, 'The Free Rider Rationale and Vertical Restraints Analysis Reconsidered' (1987) 56 *Antitrust LJ* 27. He cites most of the relevant literature. See also **Jean Tirole**, *The Theory of Industrial Organisation* (Cambridge, Mass.: MIT Press, 1988).

Other outstanding articles include: **Frederick M. Scherer**, 'The Economics of Vertical Restraints' (1983) 52 *Antitrust LJ* 687; **William S. Comanor**, 'Vertical Price Fixing, Vertical Market Restrictions, and the New Antitrust Policy' (1985) 98 *Harv. L Rev.* 983; **Oliver E. Williamson**, 'Assessing Vertical Market Restrictions: Antitrust Ramification of the

free rider argument applies—whether the licensee or dealer might reasonably be expected to increase sales by investing in tooling up, marketing, etc.[74] Barry Hawk suggests[75] that the strong point of the analyses by both the Chicago and traditional schools in the United States:

> may be preserved by giving vertical restrictions a rebuttable presumption of illegality similar to the 'quick look' applied in *Broadcast Music Inc.* v. *CBS*.[76] The presumption could be rebutted by a showing that dealer services, such as pre- and post-sales servicing of complex equipment, are not sufficiently promoted through means less restrictive than airtight vertical restraints. Such an approach is somewhat similar to that followed under Article 85, where such an arrangement would be presumed to violate Article 85, and would be exempted under Article 85(3) only upon a showing of increased efficiency.

I would prefer the analysis to be carried out under article 85(1).[77]

Transaction Cost Approach' (1979) 127 *U Penn. L Rev.* 953; **Oliver E. Williamson**, *Markets and Hierarchies: Analysis and Antitrust Implications* (New York: Free Press, 1975), 82–131; **John Flynn**, 'The "Is" and "Ought" of Vertical Restraints after *Monsanto Co.* v. *Spray-Rite Service Corp.*' (1986) 71 *Cornell L Rev.* 1095); **Eleanor Fox**, 'The Modernization of Antitrust: A New Equilibrium' (1981) 66 *Cornell L Rev.* 1140.

[74] Frederick M. Scherer points out that the free-rider argument may be taken too far. If only marginal customers require the pre-sales services, it may pay the brand owner to impose a restriction on intra-brand competition, although many customer would prefer not to pay for it: **Frederick M. Scherer**, 'The Economics of Vertical Integration' (1983) 52 *Antitrust Journal* 687,700.
   See also **Frederick M. Scherer** and **David R. Ross**, *Industrial Market Structure and Economic Performance* (3rd edn., Boston: Houghton Mufflin, 1990), 542–8, for analysis of vertical restraints. The argument about marginal buyers may apply less often to licensing agreements than to distribution arrangements, since an important sunk cost is likely to be tooling up for manufacture. Nevertheless, licensees also may have to develop a market.
[75] **Barry Hawk**, 'The American (Anti-trust) Revolution: Lessons for the EEC?' (1988) 9 *EL Rev.* 53; **Barry Hawk**, *United States, Common Market and International Antitrust: A Comparative Guide* (New York: Law & Business Inc., 1989), ii, 311.
   Articles published in Europe and advocating recognition of the free rider problem in EC law include: **Valentine Korah**, 'Goodbye, Red Label: Condemnation of Dual Pricing by Distillers' (1978) 3 *EL Rev.* 62; **Charles Baden Fuller**, 'Price Variations: The Distillers Case and Article 85 EEC' 28 *ICLQ* 128; **John Chard**, 'The Economics of the Application of Article 85 to Selective Distribution Systems' (1982) 7 *EL Rev.* 83; **Ivo Van Bael**, 'Heretical Reflections on the Basic Dogma of EEC Antitrust: Single Market Integration' (1980) 10 *Swiss Rev. Int. Comp. L* 39.
   Contrast **Michel Waelbroeck**, 'Vertical Agreements: Is the Commission Right not to Follow the Current US Policy?' (1985) 25 *Swiss Rev. Int. Comp. L* 45.
[76] 441 US 1 (1979). In *Broadcast Music* the US Sup. Ct. held that where the horizontal provisions governing a collective copyright collection society were not illegal *per se*. Hawk continues:
> 'A "truncated" rule of reason analysis was urged by the Justice Department in the *NCAA* college football case *[National Collegiate Athletic Assn.* v. *University of Oklahoma*, 468 U.S. 85]. See also the amicus brief [reproduced in XV, no. 2, *J. of Reprints for Antitrust Law and Economics*, 957] in *National Collegiate Athletic Ass'n.* v. *Board of Regents of University of Oklahoma*, US Court of Appeals, for the tenth circuit.'
[77] National courts can apply art. 85(1), but not art. 85(3). See the further reasons given in 1.1.1 above.

In the United States a licence is treated as vertical if, at the time it was negotiated, the licensee could not have entered the market without assistance from the licensor or, possibly, from some other licensor. In that situation, the licensee is not perceived to be a potential competitor, and the licence must be a method of increasing rather than restricting production.

In the EC, the Commission often used to consider a licence as horizontal once the licensee started to make products similar to those produced by the licensor, even if they were in different geographic markets.[78] This was cogently criticized by James Venit,[79] and recently the Commission has sometimes looked *ex ante*, to the time when the agreement was being negotiated.[80] Consequently, more agreements may be treated as vertical. It is not, however, easy to tell whether the licensee could have entered the market without a licence and the Commission may be wary.

Its former view, that licences should be analysed as horizontal once both licensor and licensee are using the technology to produce, explains the inclusion in the black lists of all three regulations of many provisions, such as: strong feed-back clauses;[81] customer restrictions (article 3(4));[82] quantity restrictions;[83] and restrictions on the licensee using rival technology.[84] The black list is now considerably shorter than it used to be, but some other restrictions are exempt only if notified to the Commission and it does not oppose the exemption within four months: quality specifications and tie-ins;[85] restrictions on challenging the validity of the licensed patents or the secrecy of the know-how.[86]

---

[78] This may explain some of the informal decisions mentioned in the Commission's *Tenth Report on Competition Policy* (Brussels: EC Commission, 1981), such as *Cartoux/Terrapin*, para. 129.

[79] See **Venit**, n. 43 above.

Dr Johannes, who was in charge of intellectual property licensing during the 1970s, said, n. 39 above, at 83–4:

> 'The experience of the services of the Commission shows that the overwhelming part of all patent licensing agreements is horizontal or has at least substantial horizontal effects.'

He suggested that only when an independent inventor licensed the technology was the agreement vertical. Firms operating in the same industry, even in different geographic markets, were at least potential competitors.

[80] See 1.3.1 n. 44 above. In *Delta Chemie/DDD*, cited n. 9 above and described at 3.1.5 below, at para. 24, the Commission granted an individual exemption for 20 years because the transfer of technology encourages competition. Its treatment of the agreement under art. 85(1), however, exemplifies the earlier tougher perception that the licensee might have competed with the licensor had exclusive territories not been allocated, although at para. 41 the decision states that they were not competitors. There is no reason to believe that the licensee could have competed without the benefit of the licence.

[81] Now art. 3(6).

[82] Now art. 3(4). The current version, however, applies only when the parties were already competing manufacturers before the grant of the licence.   [83] Art. 3(5).

[84] Art. 3(2).   [85] Art. 4(2)(a), analysed at 7.3.5.2 below.

[86] Art. 4(2)(b), analysed at 7.3.6.2 below.

The restrictive trade practices legislation in the United Kingdom is drafted in formalistic terms, but there are some specific exceptions for the vertical agreements most commonly made.[87]

The German experience, which has been very influential in the EC also, treats vertical agreements far more leniently than horizontal ones. It decides whether an agreement is vertical at the time of the transaction. Where the licensee could not have competed with the licensor without a licence, the licence is treated as vertical even though the parties may later compete when using the licensed technology.

A vertical agreement may be made to implement a horizontal agreement, for instance, between licensees, when the licensees persuade the licensor to restrict competition between them. In that case it is unlikely to have a pro-competitive justification and could be condemned without much market analysis without substantial risk of reducing efficiency.

Article 5 of the technology transfer regulation excludes from the ambit of the regulation many horizontal agreements, but not all. It is regretted that it does not limit many of the provisions in the black list to agreements between competitors. The Commission may have been concerned that neither it, nor national courts enforcing licensing agreements, would always be able to tell whether the parties were potential competitors.

1.4.2.1 The Commission's Green Book

Shortly before leaving the Commission, Dr Ehlermann, then Director General of the Competition Department, questioned the practice of the Commission in treating any significant restraint of conduct as having the object or effect of restricting competition. He decided that the Commission

---

[87] They may be investigated by the Monopolies and Mergers Commission under the Fair Trading Act 1973, or by the Office of Fair Trading under the Competition Act 1980, but there is no automatic sanction.

The whole policy behind the UK law is continually being reconsidered: Green Paper, *Review of Restrictive Trade Practices Policy* (Mar. 1988), Cm 331. According to ch. 4 the government considered a general prohibition, coupled with a right for third parties to sue, for 'hard core' agreements, and it lists as examples fixing prices, collusive tendering, allocating markets, restrictions on advertising, and collective refusals to supply and other collective discrimination. Only the last item is confined to horizontal agreements. Submissions have been made that the other items on the list of 'hard core' agreements should be limited to naked restraints (see 1.4.1 above) in agreements made between competitors. For the DTI's later view see its White Paper, *Opening markets: New Policy on Restrictive Trade Practices* (July 1989), Cm 727. Nevertheless, only minor changes have been made to the legislation.

In Mar. 1996 the Department of Trade and Industry issued a further consultation document, *Tackling Cartels and the Abuse of Market Power: Implementing the Government's Policy for Competition Law Reform*, again suggesting that UK law should prohibit agreements defined along the lines of art. 85(1) where trade between Member States is not affected. The time for responding has expired, a bill has been drafted and circulated for comment. Even if it is introduced, it may not be passed before the general election.

should consider all the options and, as vertical restraints are the easiest to deal with, he set up an internal committee to consider whether vertical restraints in distribution should continue to be considered anti-competitive and contrary to article 85(1).

It was expected that the Commission would publish a green book early in 1996 setting out all the options, ranging from the current practice of analysing the market only under article 85(3) to a full market analysis under article 85(1). As I write this section of the monograph in July, the green book is expected to be published after the summer holidays, but it seems to have met hostility within the Commission's secretariat, and it is no longer clear that a full range of options will be set out.

After the Commission has consulted business, Member States, and some other institutions, it hopes to adopt a white book selecting the option it will adopt. If more market analysis is performed under article 85(1), there will be far less need for the Commission to grant group or individual exemptions and national courts may be able to enforce agreements which do not restrict competition and may well increase efficiency.

1.4.3 Comparison with the US Guidelines on Intellectual Property

The Commission's case law under article 85(1) on technology licensing is in marked contrast to the US law or, more accurately, to the enforcement practice of the Department of Justice and the Federal Trade Commission.[88] In the United States, licences between firms which are not even potentially in the same market at the time the agreement is entered into are treated as vertical and seldom infringe section 1 of the Sherman Act. The analysis is broad: is the transaction likely to lead to higher prices or restricted production and, if so, does it enhance efficiency in other ways? A

---

[88] State Attorneys General have power to enforce the federal antitrust laws and recover damages on behalf of their citizens. Many have more traditional views and follow the old interventionist case law where the federal agencies would not.

Private citizens may also sue for treble damages for injury resulting from infringement of the antitrust laws. When the agencies were not pursuing vertical agreements, there were no court findings on which such actions could be based, so the plaintiff had to prove that the statutes had been infringed. Actions are, therefore, now more expensive and risky than 15 years ago. Nevertheless, some are brought, and are not directly affected by the views of the agencies.

Some of the Courts of Appeal, especially the 7th Cir. on which Richard Posner and Frank Easterbrook sit, have accepted much of the Chicago reasoning. Those two eminent thinkers have been Professors at the University of Chicago, and have been joined by Diane Wood, also from that University, but accepting more traditional views. The Supreme Court itself has moved a considerable way towards Chicago.

Currently the Department of Justice under President Clinton has been far more activist than under President Reagan, but the new guidelines on intellectual property licensing are not very much stricter than those of the 1980s.

convenient summary of the views of the Department of Justice and Federal Trade Commission, the enforcement agencies, is contained in its *Guidelines for Licensing of Intellectual Property*, published in 1995.[89]

In this section, I shall set out the US agencies' analysis of policy towards licensing generally and shall return to their policy towards specific subjects as I deal with the EC law where the analysis of policy choices is usually less clearly expressed.

The agencies state at 2.1 that they will treat exclusive intellectual property rights much like the exclusive rights of the owner of tangible property. There are differences from tangible property in that intellectual property rights are more easily misappropriated, but that can be taken into account under the traditional analysis. The benefits of technology transfer are described clearly at 2.3 of the agencies' intellectual property guidelines:

> [It] can facilitate integration of the licensed property with complementary factors of production. This integration can lead to more efficient exploitation of the intellectual property, benefitting consumers through the reduction of costs and the introduction of new products. Such arrangements increase the value of intellectual property to consumers and to the developers of the technology. By potentially increasing the expected returns from intellectual property, licensing also can increase the incentive for its creation and thus promote greater investment in research and development.

Licensing and cross licensing may avoid the problem of blocking patents:

> Field-of-use, territorial, and other limitations on intellectual property licences may serve pro-competitive ends by allowing the licensor to exploit its property as efficiently and effectively as possible. These various forms of exclusivity can be used to give a licensee an incentive to invest in the commercialization and distribution of products embodying the licensed intellectual property and to develop additional applications for the licensed property. The restrictions may do so, for example, by protecting the licensee against free-riding on the licensee's investments by other licensees or by the licensor. They may also increase the licensor's incentive to license, for example, by protecting the licensor from competition in the licensor's own technology in a market niche that it prefers to keep to itself. These benefits of licensing restrictions apply to patent, copyright, and trade secret licenses and to know-how agreements.

The concerns of the agencies and their modes of analysis are described from 3.1 of the guidelines. The agencies may be concerned if markets are divided between firms that would have competed using different technologies, or when the agreement effectively merges the r & d activities of

---

[89] Issued by the agencies on 6 Apr. 1995, and reproduced in CCH, 4 *Trade Regulation Reporter*, ¶13,132, (1995) 7 *EIPR Supp.* 3.

two of only a few entities plausibly capable of carrying out such development:

> The Agencies will not require the owner of intellectual property to create competition in its own technology. However antitrust concerns may arise when a licensing arrangement harms competition among entities that would have been actual or likely potential competitors in a relevant market in the absence of the license (entities in 'a horizontal relationship'). A restraint in a licensing arrangement may harm such competition, for example, if it facilitates market division or price fixing.

The guidelines then describe how markets will be analysed, not only the market for the final goods, but also those for the technology or even innovation markets if few firms are likely to invest in innovation. The agencies are far more concerned where there is a horizontal relationship, where the licensee was an actual or potential competitor even without the licence.

Usually restraints in licences are appraised under the rule of reason. If the restraint is unlikely to have anti-competitive effects there is no cause for concern. If anti-competitive effects are likely, the agencies will consider whether the restraint is reasonably necessary to achieve pro-competitive effects. If, however, no efficiency-enhancing integration is likely, the restraint might be declared illegal *per se* where it would be in other contexts.

Rule-of-reason analysis may be truncated if no anti-competitive effects seem likely or if the restraint is likely to increase prices or reduce production and there are no countervailing efficiencies.

4.1.1 Market structure, coordination, and foreclosure

> When a licensing arrangement affects parties in a horizontal relationship, a restraint in that arrangement may increase the risk of coordinated pricing, output restrictions or the acquisition or maintenance of market power. Harm to competition may also occur if the arrangement poses a significant risk of retarding or restricting the development of new or improved goods or processes. The potential for competitive harm depends in part on the degree of concentration in, the difficulty of entry into, and the responsiveness of supply and demand to changes in price in the relevant markets. . . .

> When the licensor and licensees are in a vertical relationship, the Agencies will analyze whether the licensing arrangement may harm competition among entities in a horizontal relationship at either the level of the licensor or the licensees, or possibly in another relevant market. Harm to competition from a restraint may occur if it anti-competitively forecloses access to, or increases competitors' costs of obtaining, important inputs, or facilitates coordination to raise price or restrict output. The risk of anti-competitively foreclosing access or increasing competitors' costs is related to the proportion of the market affected by the licensing restraint; other characteristics of the

relevant markets, such as concentration, difficulty of entry, and the responsiveness of supply and demand to changes in price in the relevant markets; and the duration of the restraint. A licensing arrangement does not foreclose competition merely because some or all of the potential licensees in an industry choose to use the licensed technology to the exclusion of other technologies. Exclusive use may be an efficient consequence of the licensed technology having the lowest cost or highest value.

Harm to competition from a restraint in a vertical licensing arrangement also may occur if a licensing restraint facilitates coordination among entities in a horizontal relationship to raise prices or reduce output in a relevant market. For example, if owners of competing technologies impose similar restraints on their licensees, the licensors may find it easier to coordinate their pricing. Similarly, licensees that are competitors may find it easier to coordinate their pricing if they are subject to common restraints in licenses with a common licensor or competing licensors. The risk of anticompetitive coordination is increased when the relevant markets are concentrated and difficult to enter. The use of similar restraints may be common and pro-competitive in an industry, however, because they contribute to efficient exploitation of the licensed property.

4.1.2 Licensing arrangements involving exclusivity

. . . Generally an exclusive license may raise antitrust concerns only if the licensees themselves, or the licensor and its licensees, are in a horizontal relationship. Examples of arrangements involving exclusive licensing that may give rise to antitrust concerns include cross-licensing by parties collectively possessing market-power . . . grant backs, and acquisitions of intellectual property rights. . . .

A second form of exclusivity, *exclusive dealing*, arises when a license prevents or restrains the licensee from licensing, selling, distributing, or using competing technologies. . . .[90] Such restraints may anti competitively fore-close access to, or increase competitors' costs of obtaining, important inputs or facilitate coordination to raise price or reduce output, but they also may have pro-competitive effects. For example, a licensing arrangement that prevents the licensee from dealing in other technologies may encourage the licensee to develop and market the licensed technology or specialized applications of that technology. . . . The Agencies will take into account such pro-competitive effects in evaluating the reasonableness of the arrangement. . . .

The antitrust principles that apply to a licensor's grant of various forms of exclusivity to and among its licensees are similar to those that apply to comparable vertical restraints outside the licensing context, such as exclusive territories and exclusive dealing. However, the fact that intellectual property

---

[90] In the EC this is called a 'no-competition' clause and it prevents the application of the group exemption. A 'best-endeavours' clause which has much the same effect, however, is permitted. See 7.3.9.2–7.3.9.5 below. The Commission has not analysed so clearly the contrasting arguments of policy.

may in some cases be misappropriated more easily that other forms of property may justify the use of some restrictions that might be anticompetitive in other contexts.

The guidelines go on to explain that if there is no anti-competitive effect, the agencies will not challenge a licence. If there is an anti-competitive effect it will consider whether the restraint is reasonably necessary to achieve pro-competitive efficiencies.

> The Agencies' comparisons of anticompetitive harms and pro-competitive efficiencies is necessarily a qualitative one. . . .
>
> The existence of practical and significantly less restrictive alternatives is relevant to a determination of whether a restraint is reasonably necessary. . . . [T]he Agencies will not engage in a search for a theoretically least restrictive alternative that is not realistic in the practical prospective business situation faced by the parties.
>
> When a restraint has, or is likely to have, an anticompetitive effect, the duration of that restraint can be an important factor in determining whether it is reasonably necessary to achieve the putative pro-competitive efficiency. . . .
>
> The evaluation of pro-competitive efficiencies, of the reasonable necessity of a restraint to achieve them, and of the duration of the restraint, may depend on the market context. A restraint that may be justified by the needs of a new entrant, for example, may not have a pro-competitive efficiency justification in different market circumstances. *Cf. US v. Jerrold Electronics Corp.*[91]

The Guidelines go on to provide a safe harbour:

4.3 Antitrust 'safety zone'
> Because licensing arrangements often promote innovation and enhance competition, the Agencies believe that an antitrust safety zone is useful in order to provide some degree of certainty and thus to encourage such activity.[92] Absent extraordinary circumstances, the Agencies will not challenge a restraint in an intellectual property licensing arrangement if (1) the restraint is not facially anticompetitive and (2) the licensor and its licensees collectively account for no more than 20 per cent of each relevant market significantly affected by the restraint.

In the case of a technology market, where market shares can hardly be usefully ascertained, the agencies enquire whether there are four independent sources of technology. The safety zone is applied at the time the conduct is in issue—not necessarily the time the agreement was concluded.

---

[91] *United States* v. *Jerold Electronics Corp.*, 187 F Supp. 545 (ED Pa. 1960), affirmed *per curiam* by the US Supreme Court, 365 US 567 (1961).

[92] The antitrust 'safety zone' does not apply to restraints that are not in a licensing arrangement, or to restraints that are in a licensing arrangements but are unrelated to the use of the licensed intellectual property.

This is not the same as the *de minimis* rule in the EC, as agreements that do not come within the safe harbour are still subject to rule-of-reason analysis, and will be upheld if the restraint is not anti-competitive in its market context or is reasonably necessary to create efficiencies. The agencies stress that few licences will be illegal under these tests. The Community Court has accepted both the need to examine a restraint in its market context[93] and the ancillary restraint doctrine,[94] but the EC Commission has taken only one decision on technology transfer since the adoption of the know-how regulation in 1989 and formerly it rejected both doctrines.

The US guidelines go on to consider the application of these general principles to specific clauses. I shall refer to these items in the guidelines when dealing with specific provisions in the Community regulation.

[93] *Delimitis*, n. 1 above.
[94] See 1.1, n. 1 and 1.4.1 above, and 2.2–2.2.5 below.

2

Case Law of the Community Court

2.1 SOME OF THE COURT'S JUDGMENTS ON THE FREE MOVEMENT OF GOODS AND SERVICES[1]

Often a licensor needs to encourage a licensee to invest in setting up a production line and developing a local market by granting it an exclusive licence and promising to keep other licensees out of the territory. This it can do only by a contractual provision, subject to article 85. The Court has developed a concept of the exhaustion of intellectual property rights. Once the product has been sold by a licensor or with its consent in one Member State, the right is said to be exhausted and cannot be relied upon to keep the product out of other Member States. One cannot advise on the amount of territorial protection that may be conferred on licensor or licensee without being familiar with the rules developed by the Court.

The principle of the free movement of goods is incorporated by articles 30 to 36 of the EC Treaty. Article 30 provides:

> Quantitative restrictions on imports and all measures having equivalent effect shall, without prejudice to the following provisions, be prohibited between Member States.

The Court has interpreted this article widely, as it constitutes a fundamental principle of Community law and not merely a rule.[2] Quantitative restrictions, or import quotas, have been held to include a nil

---

[1] There are many articles on the subject, including: **Valentine Korah**, 'The Limitation of Copyright and Patents by the Rules for the Free Movement of Goods in the European Common Market' (1982) 14 *Case Western Reserve J of Int'l. L* 7; **Christopher Bellamy** and **Graham Child**, *Common Market Law of Competition* (4th edn., London: Sweet & Maxwell, 1993), ch. 8; **Norbert Koch**, 'Article 30 and the Exercise of Industrial Property Rights to Block Imports', in **Barry Hawk** (ed.), *Thirteenth Annual Proceedings of the Fordham Corporate Law Institute* (New York: 1986), 609; **Georges Friden**, 'Recent Developments in EEC Intellectual Property Law: The Distinction between Existence and Exercise Revisited' (1989) 26 *CML Rev.* 193; **Laurence W. Gormley**, *Prohibiting Restrictions on Trade within the EEC: The Theory and Application of Articles 30–36 of the EEC Treaty* (2nd edn., Amsterdam: North Holland, 1985), 184–206.

[2] In the Treaty, the rules for free movement are headed 'Foundations' and in *Centrafarm BV* v. *Sterling Drug Inc.* (15/74), [1974] ECR 1147, [1974] 2 CMLR 480, CMR 8246, discussed at 2.1.2 below, the Court calls it a 'fundamental principle'. See also **Michel Waelbroeck**, 'The Effect of the Rome Treaty on the Exercise of National Industrial Property Rights' (1976) 21 *Antitrust Bull.* 99, 103.

quota and, when used to prevent parallel imports, intellectual property rights are measures of equivalent effect to a nil quota: no goods shall be imported without the consent of the holder.

Article 36, which derogates from the principle, has been narrowly construed.

> 36. The provisions of articles 30–34 shall not preclude prohibitions or restrictions on imports . . . *justified* on grounds of . . . the protection of industrial and commercial property. Such prohibitions or restrictions shall not, however, constitute a means of arbitrary discrimination or a disguised restriction on trade between Member States [my emphasis].

In addition, article 222 of the EC Treaty provides that:

> This Treaty shall in no way prejudice the rules of Member States governing the system of property ownership.

The system of property ownership is not the same as the rules of property ownership and was originally thought to relate only to whether a particular sector of the economy was in the private or public sector. The Court has construed this provision narrowly to avoid the partition of the Common Market.

Intellectual property rights are governed by national law. Some harmonisation has taken place under the programme for the single market and is explained shortly at 3.3 below.

More important in dividing the common market is the fact that the law of each country grants rights only for that country, although there is a single trade mark law and a single design law for Benelux—Belgium, the Netherlands, and Luxembourg. An inventor wanting protection in all Member States must take out a patent in each. The Community Patent Convention has not yet been ratified by sufficient states to come into force. If it does, it is unlikely to be much used unless the demand that the specifications and claims be translated into all eleven Community languages is dropped. More important is the European Patent Convention, which is not part of Community law, under which it has become possible to make a single application and obtain patents in all or any of the countries party to the Convention, which now includes all the Member States and a few others.[3] An international search for novelty is made by the European Patent Office and a basket of the national patents applied for is granted.

---

[3] Under the European Patent Convention, a single application may be made for patents in all the participating countries, but some Member States of the Communities have only recently joined. For this and other reasons such as cost, comparatively few inventions enjoy patent protection in every Member State of the Communities. Even under the European Convention, the fees are higher if patents for more countries are included and the technical translations into the national languages are expensive. So, often, patents are not requested for small countries with a language not used in other areas.

Each becomes valid only after the application and claims have been translated. By that time, however, it should be clearer whether it is worth investing in each translation. This Convention has been very successful and it is cheaper to use its procedure than make national applications when at least two or three national patents are required.

Since 1972, the Commission has constantly assumed that it is contrary to article 85(1) to impose export bans on licensees.[4] Before then, it was thought that by relying on the separate national laws governing intellectual property rights, a holder in several Member States might achieve the same result. It might grant a licence, for instance, under the French patent or limited to France, and rely on its national rights in Germany[5] to restrain imports there from France.

In some Member States, such as the United Kingdom, an implied licence would enable the licensee to sell where he wanted, unless the right was expressly restricted. In others, such as Germany, the doctrine of exhaustion would prevent the use of intellectual property rights to restrain the sale of goods put on the market legally, by or with the consent of the holder. The doctrine of exhaustion is not the same in all Member States, and in those where it does apply to some kinds of intellectual property rights, it may apply only within the country. Hence, before the creation of the common market, sales abroad rarely, if ever, exhausted national patent rights, although international exhaustion is more frequent under national trade mark and copyright law.

2.1.1 The Birth of the Distinction Between the Existence and Exercise of a Right

Early in the history of the common market, the Commission and Court limited the possibility of using intellectual property rights to divide the common market. In *Consten and Grundig*,[6] a German manufacturer, Grundig, appointed Consten as its exclusive distributor in France and encouraged it to develop a market for its brand of dictating machines and other products by protecting Consten from parallel imports. First, Grundig restrained its wholesalers in Germany and its distributors elsewhere from

---

[4] See *Davidson Rubber Company (Re the Agreement of)* (72/237/EEC), [1972] JO L143/31, [1972] CMLR D52, CMR 9512, described at 1.1 above and 2.1.1 below.

[5] See the early '*Voran*' potato case, 49 BGHZ 331, [1968] GRUR Int. 128, [1968] AWD 152; case in the Bundesgerichtshof, conveniently translated in part into English by **Dr Hartmut Johannes**, *Industrial Property and Copyright in European Community Law* (Leyden: A. W. Sijthoff, 1976), 113–20.

[6] *Consten/Grundig* [1964] JO 2545, [1964] CMLR 489, confirmed by the Court, *Consten SA and Grundig-Verkaufs GmbH* v. *Commission* (56 and 58/64), [1966] ECR 299, [1966] CMLR 418, CMR 8046.

exporting outside the territory of each. Secondly, it supported these export bans by arranging for Consten to own the mark 'GINT'—Grundig International—which in those days was placed on Grundig's apparatus in addition to the mark Grundig, which continued to be held by the supplier. Consten was, therefore, entitled to an injunction under French national law to restrain the import of apparatus: it would not have to consent to sales in other territories.

The contractual export bans were fairly easily condemned by the Commission, supported by the Community Court, as being contrary to article 85. It was, however, argued that Consten's right to the 'GINT' mark was protected by articles 36 and 222 of the Treaty (set out at 2.1 above and in appendix 1 below). The Commission found, however, that the ancillary agreement that enabled Consten to register the French mark infringed article 85 and ordered the parties to do nothing to impede parallel imports.[7]

In upholding the Commission's decision, the Court drew a distinction between the existence of national property rights protected by articles 36 and 222 and their exercise, which is subject to the Treaty. This has given the Court considerable flexibility. Since lawyers tend to define rights in terms of the ways they can be exercised, the distinction between existence and exercise cannot be drawn analytically. A vital difference, divided by a line that only the supreme court can draw, gives the members of that court a very wide discretion.[8] In taking advantage of this flexibility, the Court has sometimes stated, if it approves of the use being made of the rights, that the way the rights are being used goes to their substance or existence.[9]

---

[7] Grundig could have opposed the registration of a mark it had registered defensively under the Madrid Convention. Merely to assign the trade mark to a distributor does not infringe art. 85(1), when the mark is not used to impede parallel imports, *Hydrotherm Gerätebau GmbH* v. *Compact del Dott. Ing. Mario Andreoli & C. Sas* (170/83), [1984] ECR 2999, [1985] 3 CMLR 224, CMR 14,112, paras. 17–22.

[8] **Valentine Korah** (1972) 35 *MLR* 634.

[9] e.g., in *Volvo AB* v. *Erik Veng (UK) Ltd.* (238/87), [1988] ECR 6211, [1989] 4 CMLR 122, CMR 14,498, in deciding that, even if a vehicle manufacturer that held a registered design for both the vehicle as a whole and the spare parts did enjoy a dominant position within the meaning of art. 86, it would not be abusive to refuse to license independent repairers to make spare parts, the Court observed that:

'It is thus for the national legislature to determine which products are to benefit from protection, even where they form part of a unit which is already protected as such.

8. It must also be emphasised that *the right* of the proprietor of a protected design *to prevent third parties* from manufacturing and selling or importing, without its consent, products incorporating the design *constitutes the very subject—matter of his exclusive right*' [my emphasis].

This was made even clearer in the Court's judgment in *Consorzio Italiano della Componentistica di Ricambio per Autoveicoli and Maxicar* v. *Régie Nationale des Usines Renault*, (53/87), [1988] ECR 6039, [1990] 4 CMLR 265, 1 CEC 59:

11. . . . The *authority of a proprietor* of a protective right in respect of an

Where it disapproves, it can say that the question relates to their exercise, and is subject to the Treaty.

2.1.2 Exhaustion When the Product has been Sold by or with the Consent of the Holder

The next important case was *Deutsche Grammophon*,[10] where the Court rejected the use of intellectual property rights to restrain the import of goods sold abroad by the holder or with its consent. The plaintiff tried to exercise rights similar to copyright under German law to restrain the import of records it had sold in France through its French subsidiary. For the first time the Community Court relied on the rules of the free movement of goods rather than on articles 85 and 86. It interpreted article 36 and the Court's earlier distinction between the existence and exercise of intellectual property rights to permit justifications for the exercise of intellectual property rights on the basis of the specific subject matter of the particular kind of intellectual property right.

> 11. . . . Although it permits prohibitions or restrictions on the free movement of products, which are justified for the purpose of protecting industrial and commercial property, Article 36 only admits derogations from that freedom to the extent to which they are justified for the purpose of safeguarding the rights which constitute the specific subject-matter of such property.
>
> 12. If a right related to copyright is relied upon to prevent the marketing in a Member State of products distributed by the holder of the right or with his consent on the territory of another Member State on the sole ground that such distribution did not take place on the national territory, such a prohibition, which would legitimise the isolation of national markets, would be repugnant to the essential purpose of the Treaty, which is to unite the national markets into a single market.

In *Centrafarm* v. *Sterling*,[11] the Court developed this idea. Sterling held the patents for the drug 'Negram' in several Member States, and its

---

ornamental model *to oppose the manufacture by third parties*, for the purpose of sale on the internal market or export, of products incorporating the design or to prevent the import of such products manufactured without its consent in other Member States *constitutes the substance of his exclusive right*. To prevent the application of national legislation in such circumstances would therefore *be tantamount to challenging the very existence of the right*' [my emphasis].

See also *Warner Bros. and Metronome* v. *Christiansen* (158/86), [1988] ECR 2605, [1990] 3 CMLR 684, [1990] 1 CEC 33, described at 2.1.3, text to n. 25.

[10] *Deutsche Grammophon Gesellschaft mbH* v. *Metro-SB-Großmärkte GmbH & Co. KG* (78/70), [1971] ECR 487, [1971] CMLR 631, CMR 8106.

[11] *Centrafarm* v. *Sterling*, n. 2 above. One of the earliest and best critical comments was by **René Joliet**, 'Patented Articles and Free Movement of Goods within the EEC' (1975) 28 *CLP* 15, at 29–32.

subsidiaries held the local trade marks. The English subsidiary made the drug under licence and sold it to wholesalers. Centrafarm bought from a wholesaler in England and imported the drug into the Netherlands, where the prevailing price was twice as high as in the United Kingdom. This was partly because of fluctuating exchange rates and partly because the National Health Service in the United Kingdom pays for almost all drugs used for human treatment, and had powers to obtain compulsory licences. Sterling sought to restrain such imports on the basis of its Dutch patent and, concurrently, its Dutch subsidiary sought to exercise the Benelux trade mark rights in the mark 'Negram' to oppose the imports.

After stating that article 30 was a fundamental principle of Community law and should be construed widely, while article 36, which derogates from the principle, should be construed narrowly, the Court developed the concept of the 'specific subject matter' of the particular kind of right with which it was faced.

> 9. In relation to patents, *the specific subject* matter of the industrial property is the guarantee that the patentee, *to reward the creative effort of the inventor*, has the *exclusive right* to use the invention with a view to manufacturing industrial products and putting them into circulation for the first time, either directly or by the grant of licences to third parties, as well as the right to oppose infringements [my emphasis].

The Court referred to the reward to the inventor rather than to the incentives needed for investment in innovation.[12] The 'specific subject matter' of the right embraced both the reason why the law grants it and the mechanism for enabling the inventor to obtain a reward.

The Court objected to the limitation in national laws of the concept of exhaustion to sales within the country and concluded that:

> 11. Whereas an obstacle to the free movement of goods of this kind may be justified on the ground of protection of industrial property where such protection is invoked against a product coming from a Member State where it is not patentable and has been manufactured by third parties without the consent of the patentee and in cases where there exist patents, the original proprietors of which are legally and economically independent, a derogation

---

[12] See 2.1.3 below. For an incisive criticism of the Court's definition of the specific subject matter of the patent, see **René Joliet**, n. 11 above, 29–32. There are four major theories justifying patent law identified by **Fritz Machlup**, 'An Economic Review of the Patent System', *Study of the Committee on Patents, Trademarks and Copyrights of the Committee on the Judiciary, US Senate, 85th Congress, Study No. 15* (Washington, DC: 1958), 21: (1) incentive to investment and innovation; (2) reward for the same; (3) price for the publication of the specification which may be used by third parties for research purposes even during the life of the patent; and (4) the encouragement of know-how into the public domain when it will be protected only for a short time. Joliet prefers the 'monopoly-profit-incentive' thesis.

from the principle of the free movement of goods is not, however, justified where the product has been put onto the market in a legal manner, by the patentee himself or with his consent, in the Member State from which it has been imported, in particular in the case of a proprietor of parallel patents.

The Court ruled that:

> 1. The exercise, by the patentee, of the right which he enjoys under the legislation of a Member State to prohibit the sale, in that state, of a product protected by the patent which has been marketed in another member state by the patentee or with his consent is incompatible with the rules of the EEC Treaty concerning the free movement of goods within the Common Market.

This judgment considerably reduced the value of intellectual property rights. Although in some Member States the value of patent rights was reduced by the possibility of compulsory licences, as in the United Kingdom, or by the imposition of maximum price control, as in France, the ruling in *Centrafarm* v. *Sterling* does not allow a patentee to discriminate and obtain a further tribute for its innovation in those countries where intellectual property rights are more valuable.[13] The policy of some Member States to keep down the price of drugs at the expense of the pharmaceutical companies was extended to those Member States that were more concerned to encourage innovation.

This view of the Court was carried to the extreme in *Merck* v. *Stephar*.[14] Merck had sold or consented to the sale in Italy by its subsidiary of a drug called 'Moduretic'. It held patents for this in other Member States, but at the time the invention was novel, it was not possible to obtain a patent for pharmaceutical products in Italy. Although the Italian Constitutional Court later overruled the law preventing such patents being issued, no transitional provisions were ever enacted to enable inventors to apply for patents out of time for inventions that were no longer novel. Consequently, neither Merck nor its subsidiary was able to earn any monopoly profit in Italy. Nevertheless, the Court ruled that Merck could not rely on its Dutch patent to keep out of the Netherlands imports that had been sold in another Member State by it or with its consent.

The Court gave no reasons: it stated only its conclusion. It observed that a patent does not guarantee the profits of a monopolist and that, in exploiting its innovation, the holder has to take into account all the relevant circumstances. Two of these circumstances were the absence of patent protection in Italy and the possibility of parallel imports from there

---

[13] See, e.g., **Paul Demaret**, *Patents, Territorial Restrictions and EEC Law: A Legal and Economic Analysis*, IIC Studies (Weinheim: Verlag Chemie, 1978), 62–76, 86–93.

[14] *Merck & Co. Inc.* v. *Stephar BV* (187/80), [1981] ECR 2063, [1981] 3 CMLR 463, CMR 8707. See 2.1.3, text to n. 30, where the possible reversal of this judgment is described.

under the rules of free movement of goods. Since it had marketed the product in Italy, Merck must take the consequences of its choice. This was a conclusion not a reason. Without explaining why, the Court ruled that the rules on free movement prevailed over Merck's national exclusive rights: the absence of patent protection in Italy deprived it of any possibility of deriving monopoly profits on any sales there. The Court relied on the final words of paragraph 11 of the judgment in *Centrafarm* v. *Sterling*, but ignored the earlier part of the paragraph and the concept of reward in paragraph 9 (both quoted above).

The Court in *Deutsche Grammophon*[15] and *Centrafarm* v. *Sterling* did not have to decide what would happen if the product were not sold by a licensee within his own territory, but within the territory of another licensee. The wording of the judgments applies only to indirect sales in the licensee's own territory from which someone else ships it into that of another licensee. After *Deutsche Grammophon*, but before *Centrafarm* v. *Sterling*, in a judgment relating to a trade mark, which has been expressly overruled on another point,[16] *Van Zeulen* v. *Hag*,[17] the Court ruled that the holder in Belgium could not prevent direct sales from Germany to Belgium or Luxembourg. In another trade mark case, *Ideal Standard*,[18] the Court assumed that, where there was control over the products to which the mark might be applied, consent to the use of the mark in France would exhaust the German right. The Advocate General in *Pharmon* v. *Hoechst*[19] also considered that the grant of a patent licence would exhaust the rights elsewhere, but the Court did not have to decide the point and did not address it.

It seems, however, that the Commission now takes the view that the grant of the licence for one Member State does not exhaust a patent elsewhere.[20]

---

[15] N. 10 above.

[16] *Hag II—CNL Sucal* v. *Hag GF AG* (C–10/89), [1990] ECR I–3711, [1990] 3 CMLR 571, [1991] 2 CEC 457.

[17] *Hag I—Van Zuylen Frères* v. *Hag AG* (192/73) [1974] ECR 731, [1974] 2 CMLR 127, CMR 8230.

[18] *Ideal-Standard—IHT Internazionale Heiztechnik GmbH* v. *Ideal-Standard GmbH* (C–9/93), [1994] ECR I–2789, [1994] 3 CMLR 857, [1994] 2 CEC 222.

[19] *Pharmon BV* v. *Hoechst AG* (19/84), [1985] ECR 2281, [1985] 3 CMLR 775, CMR 14,206.

[20] See art. 2(1)(14) of the technology transfer reg. analysed at 7.3.14 below. The Commission permits as not infringing art. 85(1) a reservation by the licensor of his right to exercise his intellectual property rights in favour of another licensee if the first licensee should sell outside his territory. Presumably, the Commission must, therefore, consider that his rights remain valid and not exhausted under Community law by the grant of a licence, although some reward may have been obtained therefor by the licensor.

Its view is contrary to the assumption on which the patent regulation was drafted, but follows the words with which the earlier judgments describe exhaustion.

The lack of consent by the holder of the intellectual property rights was stressed by the Court in *Keurkoop* v. *Nancy Kean Gifts*.[21] The holder was allowed to exercise its right to restrain import when it had not consented to the sale in another Member State. Nancy Kean registered the design of a handbag in the Netherlands, without the original inventor objecting. The year after Nancy Kean started selling bags made to this design and imported from Taiwan, Keurkoop imported similar bags from Taiwan. Direct imports from Taiwan into the Netherlands could be restrained, as this did not affect trade between Member States contrary to article 30. Otto GmbH imported some of the bags and marketed them in Germany, whence they found their way to the Netherlands. The Community Court ruled that their import could also be opposed provided that Nancy Kean had not consented to their sale in Germany, where it held no design rights and could not have obtained them since it was not the original inventor, as required by German law.

After repeating at paragraph 18 that in the present state of Community law it is for national law to define the scope of intellectual property rights,[22] the Court observed that:

> 24. Article 36 is thus intended to emphasize that the reconciliation between the requirements of the free movement of goods and the respect to which industrial and commercial property rights are entitled must be achieved in such a way that protection is ensured for the *legitimate exercise*, in the form of prohibitions on imports which are '*justified*' within the meaning of that article, of the rights conferred by national legislation, but is refused, on the other hand, in respect of *any improper exercise* of the same rights which is of such a nature as to maintain or establish artificial partitions within the common market. The exercise of industrial and commercial property rights conferred by national legislation must consequently be restricted as far as is necessary for that reconciliation.
>
> 25. The Court has consistently held that the proprietor of an industrial or commercial property right protected by the legislation of a Member State may not rely on that legislation in order to oppose the importation of a product which has lawfully been marketed in another Member State by, or with the consent of, the proprietor of the right himself or a person legally or economically dependent on him [my emphasis].

For the first time, the Court openly distinguished between the legitimate and improper exercise of national intellectual property rights and laid the

---

[21] *Keurkoop BV* v. *Nancy Kean Gifts* (144/81), [1982] ECR 2853, [1983] 2 CMLR 47, CMR 8861.
[22] The notion goes back to *Parke, Davies & Co.* v. *Probel, Reese, Beentema-Interpharm and Centrafarm* (24/67), [1968] ECR 55, [1968] CMLR 47, CMR 8054, and has been repeated since in many cases.

ground for the exercise of its discretion in later cases on grounds more clearly based on policy.

It is difficult to reconcile this approach with the view that articles 30 and 36 are concerned with state measures, rather than with the conduct of individuals but, ever since *Deutsche Grammophon*, the Court has applied the rules for the free movement of goods to the exercise by firms of their intellectual property rights. It seems, therefore, that the state measure prohibited by Article 30 must be the national court's enforcement of the right rather than the grant of a patent or the adoption of the patent law itself.[23] After considering the possibilities of applying article 85 to any agreement, the Court ruled in *Keurkoop* v. *Nancy Kean Gifts*:

> 2. The proprietor of a right to a design acquired under the legislation of a Member State may prevent the importation of products from another Member State which are identical in appearance to the design which has been filed, provided that the products in question have not been put into circulation in the other Member States by, or with the consent of, the proprietor of a right or a person legally or economically dependent on him, that as between the natural or legal persons in question there is no kind of agreement or concerted practice in restraint of competition and finally that the respective rights of the proprietors of the right to the design in the various Member States were created independently of one another.

2.1.3 Incentives and Rewards for Innovation

In two later cases, the Court has been concerned with the issue whether the national law or its enforcement was justified by the provision of incentives to or of rewards for desirable activity. In *Thetford*,[24] the Court invoked the final sentence of article 36 to check that the national law was justified by encouraging desirable activities:

> Such prohibitions or restriction shall not, however, constitute a means of arbitrary discrimination or a disguised restriction on trade between Member States.

In *Warner* v. *Christiansen*[25] the Court rejected the traditional analysis of the case law made by Advocate General Mancini. It focused more on

[23] For a fuller analysis of the inconsistencies in the court's judgment in *Centrafarm* v. *Sterling*, see **Joliet**, n. 11 above.
[24] *Thetford Corporation* v. *Fiamma SpA* (35/87), [1988] ECR 3585, [1988] 3 CMLR 549, CMR 14,497. See also *Allen & Hanburys Ltd* v. *Generics (UK) Ltd* (434/85), [1988] ECR 1245, [1988] CMLR 701, CMR 14,446; *Commission* v. *United Kingdom of Great Britain and Northern Ireland* (C–30/90), [1992] ECR 829; and *Commission* v. *Italy* (C–235/89), [1992] 1 ECR 777, [1992] 2 CMLR 709; the national rules must not discriminate against production outside the Member State where the patent right is being exercised.
[25] *Warner Bros. and Metronome* v. *Christiansen*, n. 9 above. See also **Ewald Orf**, 'Comment' (1988) 10 *EIPR* 309.

rewards to the holder and on the custom of the industry. A dealer bought copies of a video-cassette in the United Kingdom to hire out from its shop in Denmark. In the United Kingdom at that time, copyright holders were able to oppose sale but not the hire of the cassettes, while in Denmark and in an increasing number of other countries they could oppose hire as well. The Community Court reiterated its statement in *Keurkoop* v. *Nancy Kean Gifts* (described at 2.1.2 above) that in the present state of Community law, it is for national law to define the scope of intellectual property rights. Moreover, the Court considered that the Danish law was justifiable on grounds of industrial and commercial property within the meaning of article 36 in view of the importance of the rental market and of the need for the copyright holder to earn a fair reward.[26] It ruled that the holder might restrain the hire of the films in Denmark, although it had obtained some remuneration from the cassettes when it had consented to their sale in the United Kingdom. The defendant argued in terms reminiscent of the judgment in *Merck* v. *Stephar* that Warner had chosen to market the cassette in the United Kingdom, where there was no specific right to control rental, and must take the consequences. At paragraph 18 the Court rejected this argument on the ground that it would deprive a national law permitting the control of rentals of its substance.[27]

In *Thetford*,[28] the Court recognized that the United Kingdom rule[29]

---

[26] Paras. 15 and 16. Compare *Coditel II—Coditel SA and Others* v. *Ciné-Vog Films and Others SA (No 2)* (262/81), [1982] ECR 3381, [1983] 1 CMLR 49, CMR 8862, 1.4.1 above, and 2.2.5 below, where the Court looked to the special feature of the cinematograph industry, including the system of financing production to justify its view that an exclusive territory did not, in itself, infringe art. 85(1). *Coditel* v. *Ciné Vog Film (No 1)* (62/79), [1980] ECR 881, [1981] 2 CMLR 362, CMR 8662, was decided on the rules for the free movement of services, to which the doctrine of exhaustion developed in *Centrafarm* v. *Sterling*, n. 2 above, was held not to apply, but *Warner* was concerned with goods. I am delighted that the Court based its later judgment on the commercial need for remuneration as an incentive rather than on the formalistic difference between the rules of the free movement of goods and of services.

Contrast *Ministère Public* v. *Jean Louis Tournier and Jean Verney* (*SACEM*) (395/87), [1989] ECR 2521, [1991] 4 CMLR 248, [1990] 2 CEC 815, from para. 11 where the Court referred to judgments on the free movement of goods and services respectively and stated that these were different without giving any reasons.

[27] At para. 19 the Court referred to its judgment in *Musik-Vertrieb Membran* v. *GEMA* (55 & 57/80), [1981] ECR 147, [1981] 2 CMLR 44, CMR 8670, but did not distinguish it. In *Musik-Vertrieb*, the Court did not permit GEMA to oppose the import from the UK into Germany of records which had paid a lower rate of royalty than would have been charged in Germany, because of the statutory licence to manufacture under UK law. In *Warner* v. *Christiansen*, there was no right at all in the UK to oppose rental, whereas in *Musik-Vertieb* there was a right to oppose sale unless the statutory royalty was paid.

[28] *Thetford Corporation* v. *Fiamma SpA*, n. 24 above, at paras. 19 and 20.

[29] Under the influence of the European Patent Convention, this rule in the United Kingdom has since been abrogated for new patents by the Patents Act 1977.

permitting a patent to be obtained if there was no patent specification in the last fifty years was adopted in order:

> *to foster creative activity* on the part of inventors in the interest of industry. To that end, the '50-year rule' aimed to make it possible to give a reward, in the form of the grant of a patent, even in cases in which an 'old' invention was 'rediscovered'. In such cases the United Kingdom legislation was designed to prevent the existence of a former patent specification which had never been utilized or published from constituting a ground for revoking a patent which had been validly issued.
>
> 20. Consequently, a rule such as the 50-year rule cannot be regarded as constituting a restriction on trade between Member States [my emphasis].

The reference I have italicized to the object of fostering creative activity seems to me to refer to the need for an incentive. The Court is looking *ex ante* to when incentives are relevant and to the need for investment in research. In this it goes a little further than the judgment in *Centrafarm* v. *Sterling*, which does not say why the inventor should have a reward: whether it is fair that those who take advantage of its investment should be made to pay or whether the hope of reward induces the investment. The Court in *Thetford* links the concept of fostering creative activity to the reward used in the earlier judgments. If one considers the hope of reward to be an inducement to innovation, there is no difference between the two concepts. So this rationalization is extremely welcome.

In *Merck & Co. Inc. and others* v. *Primecrown Ltd and others*[30] the Community Court was invited by the English High Court to reverse its judgment in *Merck* v. *Stephar*. When they joined the Communities, Spain and Portugal did not grant product patents for pharmaceutical products and controlled their maximum prices at levels far lower than other Member States. Under their Acts of Accession, the free movement of such products from Spain and Portugal to other Member States was postponed until three years after it became possible to obtain a product patent. Consequently, the doctrine of exaustion did not operate meanwhile.

Both countries have now acceded to the European Patent Convention, under which it is possible to obtain product patents in Spain and Portugal. The three years are up, and a large parallel trade has developed, which the Commission refused to restrain by safeguard measures.[31]

Advocate General Fennelly recommended that the Court should reverse its judgment in *Merck* v. *Stephar* on the ground that it was not possible to obtain a patent when the drug was novel, but he was not prepared to go further and suggest that there would be no exhaustion where it was

[30] And *Beecham Group plc* v. *Europharm of Worthing Ltd* (Joined Cases C–267 & C–268/95). The opinion was delivered on 6 June 1996, and judgment is expected in autumn 1996.
[31] An appeal has been brought in *Merck and others* v. *Commission* (T–60/96) against the refusal of this request [1996] OJ L122/20 ff., but it will not be decided for some time.

possible to obtain a patent in the country of export, but its value was undermined by stringent control of maximum prices.

The current problem of pharmaceutical products imported in very large quantities from Spain and Portugal is temporary. Nevertheless, it will continue for another seventeen years. It is, however, likely to be repeated each time a new Member State enters the Community from Eastern Europe. Moreover, price control in Spain and Portugal continues to be very strict. I hope the Court will go further than Advocate General Fennelly, as there is no chance to earn a monopoly profit to encourage investment in innovation where prices are controlled severely by state measures. It may, however, not be easy to say when the price control is sufficiently strict to prevent any reward being earned.

It may well be too late to argue that the state measures adopted to keep down the prices of pharmaceutical products are contrary to Article 5(2), which requires Member States to 'abstain from any measure which could jeopardise the attainment of the objectives of this Treaty', and article 3(g) which provides that the activities of the Community include 'a system ensuring that competition in the internal market is not distorted'.

2.1.4 First Sale into the Common Market

The doctrine of exhaustion is based on article 30 of the Treaty which prohibits import quotas and measures of equivalent effect only as between Member States. Consequently, the Court held in *EMI* v. *CBS*,[32] that EMI could rely on its trade mark rights in the United Kingdom, Denmark and Germany to keep out goods that had been sold in the United States with its consent. There was a world market-sharing agreement many years before the creation of the common market, and different firms came to own the mark 'Columbia' in the United States and Europe. The Court ruled that only if that agreement still had effects within the common market would it infringe article 85.

Exhaustion does, however, take place throughout the EEA by virtue of protocol 28 of the EEA agreement. Goods sold with consent in Norway cannot be kept out of Sweden by virtue of intellectual property rights and vice versa.

2.2 SOME OF THE COURT'S JUDGMENTS ON LICENSING INTELLECTUAL
PROPERTY RIGHTS

Since the exercise of intellectual property rights to protect a licensee who has to invest is not permitted, licensors often promise an exclusive territory

---

[32] (51, 86 & 96/75), [1976] ECR 811, [1976] 2 CMLR 235, CMR 8350–2.

and try to keep one licensee or his purchasers out of the territory of another.

In *Nungesser*,[33] in 1982, the Court quashed the Commission's decision that an exclusive licence necessarily infringed article 85(1). In *Maize Seed*[34] and several earlier decisions such as *Davidson Rubber*,[35] as explained at 1.1 above, the Commission had held that, where the market share of the products made by the licensed technology was significant by the time the Commission came to appraise the agreement,[36] an exclusive licence automatically infringed article 85(1). The exclusivity limited the holder's freedom to license another firm in the territory, and the hypothetical licensees might have exported to another Member State. This freedom to grant non-exclusive licences, however, was entirely unrealistic, because when the Commission explained its reasons for exempting at least manufacturing exclusivity it stated that an exclusive territory was essential if the technology was to be made available in Europe. It did not spell out the reasons for its conclusion, but might have argued that, without such protection, no licensee would have invested in creating productive capacity or creating demand for the product.

The facts in *Nungesser* were that INRA, a French research institute funded by the Ministry of Agriculture, had developed a very important variety of F1 hybrid maize seed which could be grown in the colder climate of Northern Europe. In France it exploited the new variety through French

---

[33] *Nungesser (L. C.) KG and Kurt Eisele* v. *Commission* (258/78), [1982] ECR 2015, [1983] 1 CMLR 278, CMR 8805. For simplicity of exposition, I shall not distinguish between the individual, Eisele, who was allowed to register the plant breeders' rights in Germany, and his firm, Nungesser, with which some of the later agreements were made. They constituted parts of the same undertaking.

[34] The decision from which the appeal in *Nungesser* was brought: [1978] OJ L286/23, [1978] 3 CMLR 434, CMR 10,083.

[35] *Davidson Rubber Company (Re the Agreement of)* (72/237/EEC), [1972] JO L143/31, [1972] CMLR D52, CMR 9512, and *Kabelmetal/Luchaire* [1975] OJ L222/34, [1975] 2 CMLR D40, CMR 9761. The Commission continued to exempt exclusive licences rather than to clear them until the know-how reg. was adopted, but I hoped that this was because of its desire to establish the *vires* for its group exemptions and that the Commission's practice might change once it had adopted the reg. See 3.1. Since the adoption of the know-how reg., however, only one individual decision has been adopted. See *Moosehead/Whitbread* [1990] OJ L100/32, [1991] 4 CMLR 391, [1990] 1 CEC 2127, described at 3.1.7 below.

[36] In *Burroughs/Delplanque* and *Burroughs/Geha* [1972] OJ L13/50 and 53, [1972] CMLR D67, CMR 9485, the Commission cleared an exclusive manufacturing licence because the licensor and both licensees were permitted to sell throughout the common market and were likely to do so owing to the insignificant costs of transporting plasticized carbon paper. Their combined market share at the date of the decision of all carbon paper, not just the sophisticated kind in issue, was under 10%.

In *AOIP/Beyrard* [1976] OJ L6/8, [1976] 1 CMLR D14, CMR 9801, the market share of rheostats for starter motors in France was under 8%, and the turnover just over FF 8 million, but an exclusive licence limited to France and those Member States where there was no other licensee was considered to restrict competition appreciably.

propagators and, in Germany, it enabled Nungesser's owner to register the plant breeders' rights as his own at a time when, according to paragraph 47 of the judgment, INRA could not have done so itself. The Court treated this as an exclusive licence to grow and sell in Germany. The Commission found that both the exclusive grant of the licence and various attempts to restrain parallel imports made by both parties restricted competition contrary to article 85(1), but the Court quashed the first part of the decision in part. It distinguished

> 53. . . . a so-called open exclusive licence or assignment [where] the exclusivity of the licence relates solely to the contractual relationship between the owner of the right and the licensee, whereby the owner merely undertakes not to grant other licences in respect of the same territory and not to compete himself with the licensee on that territory.

from

> an exclusive licence or assignment with absolute territorial protection, under which the parties to the contract propose, as regards the products and the territory in question, to eliminate all competition from third parties, such as parallel importers or licensees for other territories.

2.2.1 Open Exclusivity is Justified as an Incentive to Investment

At 1.4.1 above, I considered the doctrine of ancillary restraints necessary to make viable some transaction that is not in itself anti-competitive. The Court is coming to accept generally that such restraints do not necessarily infringe article 85(1), when necessary to induce investment.

In so far as the licence is open within the meaning of paragraph 53 of its judgment in *Nungesser*, the Court looked to the incentives required to induce investment. In relation to the licensor's investment, it said:

> 56. The exclusive licence which forms the subject-matter of the contested decision concerns the cultivation and marketing of hybrid maize seeds which were *developed by INRA after years of research and experimentation and were unknown to German farmers at the time when the cooperation between INRA and the applicants was taking shape*. For that reason the concern shown by the interveners as regards the protection of new technology is justified [my emphasis].

The Court referred more specifically to the need to induce the licensee's commitments and risk in growing the seed in Germany under government supervision.

> 57. In fact, in the case of a licence of breeders' rights over hybrid maize seeds newly developed in one Member State, an undertaking established in another Member State which was not certain that it would not encounter competition from other licensees for the territory granted to it, or from the

owner of the right himself, might be deterred from accepting the risk of cultivating and marketing that product; such a result would be damaging to the dissemination of a new technology and would prejudice competition in the Community between the new product and similar existing products.

The Court held that an open exclusive licence does not in itself restrict competition contrary to article 85(1). Paragraphs 56 and 57 contain one of the clearest statements by the Court that protection against free riders may be necessary to induce investment by the licensee. The opinion of the Advocate General, Madame Rozès, anticipated this development. In the case before the Court, INRA was financed by the government and might have developed the seed without the chance of the higher royalty obtainable from a licensee promised an exclusive territory. Nevertheless, the Court met the concern of the Member States that had presented arguments to the Court in the case, that licensors and licensees may need incentives. It looked in paragraph 56 to the time when the co-operation was taking shape, which is the time at which incentives are needed. Much of the argument was focused on the need to protect the licensee, and this the Court accepted clearly.

Nevertheless, it held that absolute territorial protection infringed article 85(1) and the Commission was right to refuse an exemption. The Court said:

> 77. As it is a question of seeds intended to be used by a large number of farmers for the production of maize, which is an important product for human and animal foodstuffs, *absolute territorial protection manifestly goes beyond* what is indispensable for the improvement of production or distribution or the promotion of technical progress, as is demonstrated *in the present case by the prohibition, agreed to by both parties to the agreement*, of any parallel imports of INRA maize seeds into Germany even if those seeds were bred by INRA itself and marketed in France [my emphasis].

The reasoning is inadequate. I use terms like 'of course', 'it goes without saying', or 'manifestly' when I cannot articulate a reason for my gut reaction. It may indicate that the Court was sharply divided.

2.2.2 When is an Exclusive Licence Open?

The distinction, drawn in paragraph 53 of the judgment, between an open and a closed exclusive licence is, however, not entirely clear. If a licence to produce and sell in France does not imply a licence to sell in Germany, a limited licence[37] would remain open even if the exclusive licensees were not permitted to poach in the territory of another.

---

[37] For a consideration of the concept of a limited licence, see 3.2–3.2.5 below, and art. 2(1)(14) of the technology transfer reg. considered at 7.3.14 below. At 2.1.2 above, I stated

The second part of paragraph 53, however, seems to suggest that if the licence does not extend throughout the common market, it confers absolute territorial protection.[38] In theory, Nungesser did not enjoy absolute territorial protection since merchants could not have been prevented from buying in France and selling in Germany owing to the rules for the free movement of goods, considered at 2.1.2–2.1.3 above. Once a French grower sold the seed to a merchant, the intellectual property rights would have been exhausted throughout the Common Market and neither INRA nor Nungesser would have been able to sue successfully to restrain commercial imports into West Germany.

Although this argument was put by the British Government, the Court did not accept it. In fact, Nungesser had managed to restrain two merchants from importing the seed from France during the early 1970s when Community law was undeveloped. INRA had also told the French growers not to export.

It seems that the case may establish any one of three propositions: (1) a licence may be closed when it is limited to part of the common market; (2) it may be closed in that case only if one or other of the parties attempts to restrain cross-frontier trade; and (3) it may be closed when the licence is limited and both parties attempt to restrain cross-frontier trade.

The Commission adopted the first proposition in *Boussois/Interpane*.[39] The second proposition would have the disadvantage that one party could, by its unilateral act, cause clauses in a licence that protect the other party to become unenforceable. The third version is still arguable in view of the Court's statement at the end of paragraph 77. A licensee wishing the exclusive territory to remain valid would be unwise to attempt to restrain imports by a court action, which has been doomed to failure at least since *Centrafarm* v. *Sterling*.[40]

2.2.3 The Meaning and Relevance of Novelty

The Court mentioned novelty in paragraphs 56 and 57 of its judgment, but in rather different terms. Is it the product or the technology that should be new, and should it be new altogether, or merely in the territory of the

---

that the grant of a licence does not exhaust a patent right and that the Commission's view now is that one may exercise a patent to keep out direct sales by a licensee outside his territory. This may not be important when the product passes through dealers, as one could not keep out indirect sales by a purchaser within the licensee's territory, but it might be important for tailor-made products.

[38] A phrase used by both Commission and Court in *Consten and Grundig*, n. 6 above, before the doctrine of exhaustion was developed.

[39] [1987] OJ L50/30, [1988] 4 CMLR 124, CMR 10,859, described at 3.1.3 below.

[40] Cited n. 2 above and described at 2.1.2 above.

licensee? In paragraph 56, quoted at 2.2 above, the Court referred not only to INRA's research, which was done in France, be also to the fact that the seeds 'were unknown to German farmers at the time when the co-operation . . . was taking shape'.[41] At paragraph 57, it speaks of 'breeders' rights over hybrid maize seeds newly developed in one Member State'.

In my view, when assessing whether the investment of licensor or licensee justifies 'open exclusivity' one should look to whether the technology or the product is new. The judgment in *Nungesser*, described at 2.2 above, was based on the need to induce investment and, even if the product is not new, investment may be required to produce a better product or as good a product more cheaply. Moreover, partly from the lack of clarity about the criterion of novelty, one may argue that the Court was not stating that either product or technology need be new. The novelty and importance of the new variety were among the factors that caused the Court to consider that an exclusive territory that might be needed to induce investment was not necessarily prohibited by article 85 (1).

In its decisions adopted to establish the *vires* for the block exemption for know-how licences, however, the Commission has stated that where the product is not new, either because there are rival technologies for making something similar as in *Rich Products/Jus-Rol*,[42] or where the licensee first acted as distributor for the product before sufficient demand was induced to warrant manufacture as in *Delta Chemie/DDD*,[43] an open exclusive territory does infringe article 85(1) and requires exemption. At 3.1 below, I suggest that the Commission was using these agreements to demonstrate the need for a group exemption. Had the licences been cleared this would not have been done. The Commission may have changed its attitude in unpublished comfort letters, but there is no evidence of this save the dearth of formal decisions after the adoption of the know-how regulation.

2.2.4 Does the Judgment in *Nungesser* Apply Only to Plant Breeders' Rights?

After pointing to the need for incentives to induce investment by both parties, as described at 2.2.1 above, the Court concluded:

> 58. Having regard to the *specific nature of the products in question*, the Court concludes that, in a case such as the present, the grant of an open

---

[41] This is an example of the Court looking *ex ante* to the time when the co-operation started, rather than to the time when the Commission's decision was made. See 1.3.1 above.

[42] *Rich Products/Jus-Rol* [1988] OJ L69/21, [1988] 4 CMLR 527, CMR 10,956, described at 3.1.4 below.

[43] *Delta Chemie/DDD*, [1988] OJ L309/34, [1989] 4 CMLR 535, [1989] 1 CEC 2254, described at 3.1.5 below.

exclusive licence, that is to say a licence which does not affect the position of third parties such as parallel importers and licensees for other territories, is not in itself incompatible with Article 85(1) of the Treaty [my emphasis].

What is specific about the product? Again there are three possibilities: first, the Court may have distinguished plant breeders' rights from patents although, earlier in the judgment,[44] it had held that breeders' rights were not so different as to be outside article 85(1) altogether; secondly, it may have indicated that plants are different from many other products in that, under the international conventions requiring the protection of breeders' rights, the variety must be distinct, uniform, stable and useful: it is alive, delicate, must be grown afresh each year from the basic seed and not from hybrids, and poor seed must be discarded; thirdly, the Court may have referred to the fact that there were many other varieties of maize seed being grown commercially and that the sale of the INRA varieties was subject to competition from them.

The headings in the judgment relate to exclusive territories in general, and not to plant breeders' rights in particular, but the text is carefully limited to plant breeders' rights. I understand that the headings would have been in the text considered by the judges when deliberating on the text of the judgment. Much is left uncertain in the judgment, probably because of the difficulty of obtaining agreement between the judges. The Commission accepts in the recitals to all three technology licensing regulations that the idea of open exclusivity applies to technology licensing, although, for reasons to be considered at 3.1 below, it has construed the concept so narrowly in its case law as virtually to eliminate it.

The reasoning in paragraphs 56 and 57 of the judgment in *Nungesser* is clearly based on providing incentives to both licensor and licensee, so in my view the judgment applies whenever this is necessary to induce useful investment. This may be a difficult test on which to advise. So, for the reasons given in 2.2.5 below, the precedent may apply to all technology licences, even if investment is not very great. It seems that Nungesser's investment was not very great.

2.2.5 Conclusion on *Nungesser* and Later Cases

Paragraph 77 of the judgment, considered at 2.2.1 above, about closed licences being manifestly too restrictive to be exempted does not seem consistent with the view of the Court in paragraphs 56 and 57 under article 85(1), also quoted at 2.2.1 above, or with later cases on the compatibility with article 85(1) of ancillary clauses necessary to make viable some transaction that is not, itself, anti-competitive. The inconsistencies and

---

[44] At para. 41.

vagueness in the judgment indicate that there must have been profound disagreement between the judges. What is very important is that the Court accepted that some element of exclusivity escapes the prohibition of article 85(1) when it is required to induce investment. Not all restrictions of conduct restrict competition.

There are major uncertainties about the extent of the judgment, but in recitals to both the patent and know-how regulations the Commission treats it as applying to technology licences generally,[45] although in its later decisions[46] it has construed the judgment virtually to nothing. As will be explained at 3.1.4 below, it is thought that this is largely because the Commission seems to believe that it has power to grant a block exemption only when it has gained experience in granting individual exemptions. I hoped that once its block exemption for know-how licences had been adopted, it would apply the Court's judgment more broadly, but the only decision adopted since has been *Moosehead /Whitbread*.[47]

I described some of the later judgments such as *Coditel (II)*[48] and *Erauw-Jacquéry*[49] at 1.4.1 above. Again, it is not entirely clear how far they apply generally to technology licensing. The first was concerned with copyright in performing rights and the second with plant breeders' rights in basic seed, but the Court's reliance on trade practice and a fair reward in *Coditel II* and on the need to induce investment in *Erauw-Jacquéry* was quite general, and it may be argued that the judgments apply also to technology licences although this is not accepted by officials in the Commission. This is particularly important if the licence cannot easily be brought within a group exemption as in the case of software that is not ancillary to the other technology.

Some lawyers do not consider the difference between incentives and rewards important. They believe that the hope of a reward is an incentive. The problem with the reward theory, adopted in *Centrafarm* v. *Sterling*,[50]

---

[45] Recital 11 of the patent reg., recital 6 of the know-how reg., and recital 10 of the technology transfer reg.

[46] Discussed from 3.1.4–3.1.5 below. It gave a very narrow meaning to the concept of novelty in *Delta Chemie/DDD*, n. 43 above and described 3.1.5 below: the licensee had previously been selling the licensed product as a distributor, so the products were treated as no longer being new. In *Rich Products/Jus-Rol*, n. 42 above and described at 3.1.4 below, there were other techniques for freezing yeast dough; it seems they were not as good, but the technology was held not to be new. This is difficult to reconcile with *Nungesser*, n. 33 above, which treated the variety or the technology as new, although maize seed has been known for generations.

In most of the batch of decisions adopted shortly before the know-how reg. there was probably a need for inducing innovation by the licensee by the grant of an exclusive territory; but the agreements were exempted, not cleared.

[47] Cited n. 35 above, and described at 3.1.7 below.   [48] N. 26 above.

[49] *Erauw-Jacquéry (Louis) Sprl* v. *La Hesbignonne Société Coopérative* (27/87), [1988] ECR 1919, [1988] 4 CMLR 576, [1989] 2 CEC 637.

[50] Cited n. 2 above and described 2.1.2 above.

is that it leads to national courts having to decide what reward is fair after the event, as the Community Court seems to advocate in *Coditel II* [51] and *Volvo*. [52]

The incentive theory requires that the person planning to invest must know in advance that if the investment turns out well, it will be able to appropriate the benefits it brings. On this basis, there must be some very large rewards for the winners to balance the possibility of the investment proving commercially useless. There is no need for anyone to control the profit the successful investor makes. The market rewards the successful and reduces the profit of those less successful at producing what is demanded. In *Nungesser, Thetford*, [53] and *Erauw-Jacquéry*, the Court has stressed the need for incentives. In *Centrafarm* v. *Sterling, Coditel II, Renault*, [54] *Volvo*, and *Warner* v. *Christiansen*, [55] the stress has been more on fair rewards. It is hoped that the latter should be interpreted as necessary to provide incentives.

2.3 REFUSALS TO LICENSE AND ARTICLE 86

Article 86 forbids, as incompatible with the common market, the abusive exploitation of a dominant position within the common market or a substantial part of it. The Court has construed this to forbid not only exploiting customers and suppliers, but also exclusionary conduct that does not amount to normal competition based on performance. Is the retention by a dominant firm of an exclusive right gained by competition in innovation based on performance?

Patents and other intellectual property rights do not necessarily confer a dominant position on the holder. In *Parke, Davies* [56] and *Deutsche Grammophon*, [57] the Court accepted the view of the Advocate General that it depends on whether there are good substitutes. Sometimes it is possible to invent around a patent, so it is not a very effective barrier to entry, and a patent that is about to expire will not exclude new entrants for long.

Where, however, the holder does enjoy a dominant position, the question arises whether he is required to grant a licence to anyone prepared to pay a reasonable royalty. In *Renault* [58] and *Volvo* [59] the automobile brand owners refused to grant a design or copyright licence to independent repairers to make spare parts. The Court in both judgments

---

[51] Cited n. 26 above and described at 1.4.1 above.
[52] *Volvo AB* v. *Erik Veng (UK) Ltd*, cited n. 9 above and described at 1.4.1 above.
[53] N. 24 above; see 2.1.3 above.   [54] N. 9 above; see 1.4.1 above.
[55] N. 9 above; see 2.1.3 above.   [56] N. 22 above.
[57] N. 10 above; see 2.1.2 above.   [58] N. 9 above; see 1.4.1 above.
[59] N. 9 above; see 1.4.1 above.

repeated its view that it was for national law to determine the scope of national intellectual property law. It assumed, without deciding the point, that the holder might be dominant, but ruled that the exclusive right to exploit the design was part of the specific subject matter of the intellectual property right and that, even if dominant, the holder was not required to grant a licence even on reasonable terms. The Court went on in both judgments to hold, however, that it might be abusive both to refuse a licence to independent repairers and to refuse to supply them with the spare parts, to charge an unreasonably high price for the parts, or to cease making the parts when there was a considerable number of vehicles of that model still on the road.

We thought we could safely advise holders of intellectual property rights that there was no duty to license, although the three qualifications to this statement were not entirely clear. On the one hand they might not be exhaustive, while on the other they seem to have been *obiter*.[60] The criteria on which reasonable prices for the products are to be determined are not established, although Advocate General Mischo considered that part of the costs of design might be recovered from the spare parts.

Now, a further qualification of uncertain ambit has to be made to the view that there is no duty to license. In *Magill*,[61] the three television companies whose programmes could be seen in Ireland and Northern Ireland held copyright in their listings and each used it to restrain Magill from compiling a comprehensive guide covering all three stations.

Lawyers from many countries, not only from civil law jurisdictions who tend to think mainly of moral rights that are being protected, would consider that copyright protection for information is improper. The draft database directive permitted the use of information not available elsewhere that is obtained from a database.[61a] The Court, however, did not have the courage to say that there are limits to the intellectual property rights that national law may protect.

The Court said:

> 49. . . . Further, the exclusive right of reproduction forms part of the author's rights, so that refusal to grant a licence, even if it is the act of an undertaking holding a dominant position, cannot in itself constitute abuse of a dominant position (judgment in Case 238/87 *Volvo*, cited above, paragraphs 7 and 8).

---

[60] There may be no formal distinction in Community law between the *ratio decidendi* of a judgment and *obiter dicta*, but when the examples were raised in argument, there was no one with any interest in attacking them. See, however, para. 50 of the judgment in *Magill—Radio Telefis Eireann (RTE) and Independent Television Publications Ltd (ITP)* v. *Commission* (C–241 & 242/91P), [1995] ECR I–743, [1995] 4 CMLR 718, [1995] 1 CEC 400, quoted at 1.4.1 above.   [61] *Magill*, n. 60, discussed at 1.4.1 above.

[61a] This provision was deleted in the final draft, perhaps on the ground that *Magill* would have the same effect, perhaps because of intense lobbying.

50. However, it is also clear from that judgment (paragraph 9) that the exercise of an exclusive right by the proprietor may, in exceptional circumstances, involve abusive conduct.

The Court went on to note the circumstances that had been taken into account by the Court of First Instance in finding the use of copyright abusive: (1) there was no actual or potential substitute for a comprehensive guide;[62] (2) there was a specific, constant, and regular potential demand for a comprehensive guide; (3) the refusal to provide the basic information over which each TV company was dominant prevented the appearance of a new product which the appellants did not offer and for which there was potential demand; (4) there was no justification for such a refusal; and (5) the appellants were reserving to themselves the secondary market for weekly television guides, since they denied access to the basic information which is needed to compile such a guide. In those circumstances, the Court confirmed that this was abusive. It accepted that a reasonable royalty might be charged, but it gave no criteria for its determination. Under national legislation where compulsory licences are required, the royalty is not the cost of the opportunity to the holder: the holder usually earns less than he would were he to produce himself.

At a conference on international intellectual property law organized at Fordham University in April 1995,[63] where the judgment in *Magill* was discussed, Christopher Bright raised the question whether the holder should be entitled only to the cost of granting the licence or also to some reward for the original investment. He thought that the Commission would want to avoid such a decision and might require the parties to arbitrate. Enrique Gonzalez Diaz, who works for the Commission's legal service, speaking in his personal capacity, suggested that the matter might be left to national courts or competition authorities. It may be some time before any answer is forthcoming.

The judgment goes very far if widely interpreted. On a literal view the holder of an improvement patent for a new product would be entitled, on payment of a reasonable royalty, to a licence from a dominant firm that holds the basic patent. The five conditions prevail. On the other hand, to require a licence might reduce the incentive to invest in the r & d which led to the original patented invention. If, contrary to a widely accepted view, the judgment were to apply to improvement patents, care should be taken to insist on a cross-licence to improvements at the time the compulsory licence is granted.

Many experts, including officials in the Court and the Commission's

---

[62] The Court did not accept that the three individual guides amounted to an actual substitute.

[63] To be published in 1996 by Juris Publications (**Prof. Hugh C. Hansen** (ed.)).

legal service, however, are now saying that in their personal views the judgment should be limited to unmeritorious kinds of intellectual property rights and does not apply to patents that may be based on extensive research expenditure. This may lead to great uncertainty, as every unusual intellectual property rule may be challenged. Unfortunately, the only member of the Community Court with substantial expertise in intellectual property, Judge René Joliet, died shortly after the judgment was delivered and did not participate in the final deliberations. How far the obligation to license extends remains uncertain. The Court's press release stressed that the circumstances were exceptional.

3

The Commission's Contribution

The Commission took several decisions[1] relating to know-how licences in order to gain experience to prepare its group exemption for know-how in 1988. In almost all of them, any element of territorial exclusivity and associated restrictions on manufacture or sale in other Member States was exempted rather than cleared. Such provisions were found to restrict competition without reference to any analysis of the market, save that the market share at the date of the decision was not *de minimis*.[2] Nor did the Commission consider whether open exclusivity was necessary to induce investment. Perhaps, as in the case of the franchising regulation,[3] the Commission thought that it was necessary to exempt some know-how agreements in order that the contemplated regulation should come within the *vires* of regulation 19/65, recital 4 of which states:

> Whereas it should be laid down under what conditions the Commission, in close and constant liaison with the competent authorities of the Member States, may exercise such powers *after sufficient experience has been gained in the light of individual decisions* and it becomes possible to define categories of agreements and concerted practices in respect of which the conditions of Article 85(3) may be considered as being fulfilled [my emphasis].

I regret that the Commission thinks that it can gain the experience only from exemptions and not from negative clearances. See 3.1.5, text to note 36 below.

---

[1] The Commission's decisions on patent licences are now very old—before the Court's judgments in *Nungesser (L. C.) KG and Kurt Eisele* v. *Commission* (258/78), [1982] ECR 2015, [1983] 1 CMLR 278, CMR 8805, described at 2.2–2.2.5 above; *Coditel II—Coditel SA and Others* v. *Ciné-Vog Films and Others SA* (262/81), [1982] ECR 3381, [1983] 1 CMLR 49, CMR 8862; and *Erauw-Jacquéry (Louis) Sprl* v. *La Hesbignonne Société Coopérative* (27/87), [1988] ECR 1919, [1988] 4 CMLR 576, [1989] 2 CEC 637, both discussed at 1.4.1 above. They were mostly analysed in my earlier monograph,**Valentine Korah**, *Patent Licensing and the EEC Competition Rules—Regulation 2349/84* (Oxford: ESC Publishing, 1985).

[2] Following *Davidson Rubber Company (Re the Agreement of)* (72/237/EEC), [1972] JC L143/31, [1972] CMLR D52, CMR 9512, described at 1.1 above.

[3] Reg. 4087/88 [1988] OJ L359/46.

3.1.1 *Mitchell Cotts/Sofiltra*

In *Mitchell Cotts/Sofiltra*[4] provisions other than territorial protection, which all three technology licensing regulations stated were not likely to restrict competition, were held in quite general terms not to do so. Sofiltra entered into a joint venture with Mitchell Cotts and granted a know-how licence to the joint venture.[5] The Commission stated at paragraph 21 that certain obligations were not appreciable restrictions of competition: the licensee's duty to keep the know-how secret, not to grant sub-licences, to pass back improvements on a non-exclusive basis without payment, and the licensor's obligation to pass on any improvements to the licensee. These were cleared, virtually *per se.*

The Commission went on to clear, 'in the specific circumstances of the case', the licensee's duty not to deal in competing products. The parties had invested considerable sums to create new production capacity, and a licence to a joint venture is a good way of disseminating new technology and increasing production.[6] The Commission concluded that it was desirable that the joint venture should concentrate its efforts on the particular technology. These circumstances prevail rather generally, so there were hopes that the block exemption might clear such a clause, but it has in fact been blacklisted by article 3(2) of the technology transfer regulation, although a promise to use one's best endeavours to exploit the licensed technology and to pay minimum royalties is whitelisted.[7] These give some assurance that the know-how will be exploited and not be used for the competing products.[8]

---

[4] [1987] OJ L41/31, [1988] 4 CMLR 111, CMR 10,852.

[5] Although the licence was part of the agreement setting up the joint venture, the Commission stated that Mitchell Cotts could not have manufactured the sophisticated filters without help from Sofiltra. So, had the know-how or technology transfer reg. then been in force, it is thought that its application would not have been prevented by art. 5(1)(2), analysed at 5.1.1.2 below. The Commission did state that at the level of distribution and sales, the joint venture and Sofiltra were competitors. This, however, would have been to look at the transaction *ex post* and not *ex ante.* The Commission is becoming increasingly aware of the need to look *ex ante*: 1.3.1 above. Moreover, from the factual part of the decision, it looks as if a French company, Sofiltra, might have had great difficulty penetrating the nuclear power-supply industry in the United Kingdom, then publicly owned. They had the same products to sell, but I doubt whether they could compete in the same geographic markets.

[6] See *Optical Fibres* [1986] OJ L236/30, CMR 10,813; **Valentine Korah**, 'Comment, Critical Comments on the Commission's Recent Decisions Exempting Joint Ventures to Exploit Research that Needs Further Development' (1987) 12 *EL Rev.* 18.

[7] See 7.3.9–7.3.9.5 below, and compare the analysis in sect. 4.1.2 of the US guidelines on intellectual property, quoted at 1.4.3 above. A best-endeavours clause in an agreement between competitors in a concentrated market protected by high entry barriers could amount to a way of ensuring that other technology was never used. Those conditions, however, did not prevail in *Mitchell Cotts.*

[8] See *Delta Chemie/DDD* [1988] OJ L309/34, [1989] 4 CMLR 535, 1 CEC 2254, analysed at 3.1.5 below.

3.1.2 *Campari*

Campari[9] was decided as early as 1978. The Italian holder exploited its mark and know-how in Member States other than Italy by granting an exclusive trade mark licence. Such a transaction might these days be called a 'production franchise' or an 'exclusive trade mark licence'. The licensees were provided with the mixture of secret herbs and instructed how to make the bitters by infusing the herbs in local wines. They then placed the liquid into bottles got up and marked in the Campari way. The recipe for the wines may have been sufficiently substantial to qualify as 'know-how' for the purpose of the know-how and technology transfer regulations, although the know-how relating to the secret herbs was not transferred.

The Commission held that the exclusive licences and the restraints on actively seeking sales outside the territory restricted competition contrary to article 85(1), as did the restraint on handling competing products and the obligation to supply only the original Italian product and not the locally made variety to bodies with duty-free facilities.

The Commission cleared the restriction of the production licence to plants capable of guaranteeing the quality of the product on the ground that it did not go beyond a legitimate objective of quality control by the holder of the licensed mark. On the same ground it cleared an obligation for the licensee to follow the licensor's instructions and buy the secret raw materials from the licensor.

The Commission also cleared the obligation of the licensees to maintain continuous contact with customers and to spend a standard minimum sum on advertising, since the amount was not so high as to prevent the licensees from engaging in other activities or from carrying their own advertising as well. It implied that if the obligation had been more onerous, it might have required exemption. The duty to promote is permitted by article 2(3) of the exclusive distribution regulation and described as an 'obligation' rather than as a 'restriction of competition', implying that in distribution agreements it should be cleared. Nevertheless, in *Metro I*,[10] the Court treated a restriction on selling to unauthorized dealers, when authorized

---

[9] *Re the Agreement of David Campari Milano SpA* (78/253/EEC), [1978] OJ L70/69, [1978] 2 CMLR 397, CMR 10,035.

[10] *Metro* v. *Commission* (26/76) [1977] ECR 1875, [1978] 2 CMLR 1, CMR 8435, paras. 27, 39–40. Contrast the Commission's decision in *Villeroy and Boch* [1985] OJ L376/15, [1988] 4 CMLR 461, CMR 10,758, para. 30, where the market was competitive and the relationship vertical. The dealers' obligations to promote were cleared. In that case, however, dealers were expected to sell competing brands, so the duty to promote cannot have foreclosed other suppliers. Unfortunately, the Commission has returned to a more interventionist approach in its later decision on selective distribution. See **Valentine Korah** and **Warwick Rothnie**, *Exclusive Distribution and the EEC Competition Rules—Regulations 1983/83 and 1984/83* (2nd edn., London: Sweet & Maxwell, 1992), 266, and ch. 10 generally.

dealers were subject to a duty to promote, as a quantitative restriction that required exemption.

In *Campari*, the Commission exempted the exclusive territories on the ground that the protection enabled a licensee to make a sufficient return on its investment. Campari had induced each licensee to increase production capacity and invest in improving its distribution network. The Commission said that the obligation accepted by a technology licensee not to handle competing products might constitute a barrier to technical and economic progress by preventing the licensee from taking an interest in other techniques and products. In this case, however, the aim was to decentralize production and rationalize distribution and the Commission exempted the clause by analogy to regulation 67/67.[11]

The Commission also exempted the restriction on active sales outside the territory on the ground that it helps to concentrate the efforts of each licensee in its territory. In the United States, unless the supplier had significant market power such a restriction would be justified as being necessary to protect the investment by neighbouring licensees in establishing production and making a market.

The restriction on selling the locally made product to duty-free facilities was justified on the ground that they served a clientèle that travels and might wish to obtain the same product made by the licensor from Italian wines throughout Europe.[12]

When the exemption came to an end in 1986, Campari had changed its strategy. It produced the bitters itself for the whole common market. In some Member States its subsidiaries carried out the distribution and, in others, it entered into exclusive distribution agreements. The Commission

---

[11] The predecessor of regs. 1983/83 and 1984/83 granting group exemptions for exclusive distribution and purchasing agreements. The Commission seems more prepared to exempt a restriction on the licensee competing in what it treats as distribution agreements than in technology licences. See Art. 2(2)(a) of reg. 1983/83, analysed in **Korah** and **Rothnie**, n. 10 above, 103–4.

Restrictions on competing are blacklisted in the three group exemptions for technology licences. This may be a relic of its view that most licences are horizontal once licensor and licensee are producing by means of the technology, even if the licensee could not have produced without taking a licence. See 1.4.2 above. This theory may also explain the idea implied in *Campari*, that an onerous obligation to promote goods made under a production franchise may require exemption.

There has been a progressive shift in policy since the patent reg. was adopted, and now a best endeavours clause, which may have much the same effect, is permitted by art. 2(1)(17) and (18) of the technology transfer reg., analysed at 7.3.9–7.3.9.5 below.

[12] The Commission considered that different manufacturers could not obtain a uniform taste, but does not say why they could not. One of the main differences may have been the alcoholic content, a difference due in part to national tastes and in part to differences in national law before the judgment in *Cassis de Dijon—Rewe-Zentral AG v Bundesmonopol-verwaltung für Branntwein* (120/78), [1979] ECR 649, [1979] 3 CMLR 494. Another may have been that the trade mark licensees used to use local wines.

issued a press release[13] stating that the new network was compatible with the competition rules. This network does not seem to be any more competitive than the old arrangements that required exemption. Indeed, competition in production has been eliminated and transport costs may have been increased. The exclusive distribution agreements may well have come within the group exemption for exclusive distribution agreements, regulation 1983/83, which permits an obligation on the re-seller not to pursue an active sales policy outside its territory, but this is not stated. If Campari changed its strategy because it considered it better to manufacture itself, the change may have increased efficiency. It is possible, however, that it changed its strategy in order to avoid the Commission's tight control over licensing agreements, and this may have added to its costs and to the price of its product as well as eliminating local variations in taste, and so have reduced welfare within the common market.[14]

Even if in the original agreement the trade mark licence was ancillary to that of the know-how, and the contract would have qualified as a know-how licence,[15] any period of territorial protection exempted under the know-how regulation must have expired, but Campari is still able to grant an exclusive territory and to impose on its distributors a restriction on active sales under article 2(2)(c) of regulation 1983/83. It is surprising that a longer period of territorial protection from active sales is permitted for distribution agreements than for licences, although distributors do not have to invest in developing production facilities. There are still traces of the Commission's old view that patent licences are horizontal once parties have started to manufacture, whereas most exclusive distribution agreements are vertical.[16]

For many years, there was no decision other than *Campari* relating to know-how licences[17] and it is at least doubtful whether the instructions for

---

[13] IP [88] 594. The Commission issued a comfort letter that seems to have been of the neutral kind, not saying whether the agreement merits exemption, is cleared as not contrary to art. 85(1), or benefits from a group exemption. Campari was asked to supply information about agreements relating to other Member States, and to make certain amendments which the Commission does not specify.

[14] See **Oliver E. Williamson**, 'Markets and Hierarchies: Some Elementary Considerations' (1973) 63 *American Economic Review* 316. His views were developed further in *Antitrust Economics: Mergers, Contracting, and Strategic Behaviour* (Oxford: Basil Blackwell, 1987).

[15] Arts. 1(1) and 5(1)(4) of the know-how reg. prevented it from applying to trade mark licences unless the mark was ancillary to the know-how. This may no longer be the position in view of the definition of 'ancillary' in art. 10(15) of the technology transfer reg. See 5.1.4 below.

[16] See **James Venit**, 'In the Wake of *Windsurfing*: Patent Licensing in the Common Market' in **Hawk** (ed.), *Thirteenth Annual Proceedings of the Fordham Corporate Law Institute* (New York: 1986).

[17] Apart from *Reuter/BASF* [1976] OJ L254/40, [1976] 2 CMLR D44, CMR 9862, mentioned at 1.3.2 n. 46 above, and *Schlegel/CPIO* [1983] OJ L351/20, [1984] 2 CMLR 179,

infusing the herbs in wine were sufficiently complex to qualify as substantial, secret and recorded know-how. One decision in *Moosehead/ Whitbread*[18] was, however, adopted after the know-how regulation was adopted.

3.1.3 *Boussois/Interpane*[19]

Boussois/Interpane[20] is the first of the decisions on know-how licences adopted to provide expertise for the formulation of the group exemption for know-how licences.[21] Interpane licensed Boussois to use its know-how for applying thermal insulating coatings to glass for double glazing in facilities provided in France by the licensor. The product was held to be new, and more efficient than normal double glazing, but competed with it and with triple glazing. The coating actually exploited was protected by patents in seven Member States, but licensed know-how relating to methods for applying two other coatings were not so protected anywhere. They were not yet being exploited.

The Commission accepted the parties' view that the licence to manufacture in the plant erected by Interpane in France impliedly restrained manufacture in other Member States.[22] Moreover, Interpane promised not itself to manufacture in France for two years and that other licensees would be restrained from manufacturing or selling in France for five years. These elements of exclusivity were found to restrict competition. Boussois' licence to sell in areas where no exclusive licence had been granted was subject to the possibility of further exclusive licences being granted, and this proviso, too, was found to be caught by the prohibition of article 85(1).

The Commission did not treat the restrictions on direct sales by other licensees as open exclusive licences not necessarily caught by article 85(1),

---

CMR 10,545. The Commission cleared a 7-year licence for the non-exclusive use of know-how. There were no restrictions on competition at all in that agreement. The Commission exempted, however, a parallel agreement under which the licensee agreed to obtain a component from the licensor for 5 years. The Commission considered that 5 years went beyond normal long term sales agreements. This view in relation to art. 85(1) is difficult to reconcile with its statement under art. 85(3) that such long-term agreements are normal practice in the automobile industry and essential when series production of vehicles is being planned. It was justified on grounds both of quality control and of rationalizing production at both levels.

[18] [1990] OJ L100/32, [1991] 4 CMLR 391, [1990] 1 CEC 2127, described at 3.1.7 below.
[19] See also 1.2, text to n. 34 above, and 6.3 below.
[20] [1987] OJ L50/30, [1988] 4 CMLR 124, CMR 10,859.
[21] See 3.1 above.
[22] For the Court's judgments on the free movement of goods see 2.1–2.1.3, above.

under the Court's decision in *Nungesser*,[23] even though no restrictions were imposed, as they were in *Nungesser*, on purchasers from other licensees. The Commission now seems to take the view that the grant of a licence in itself does not exhaust a patent right,[24] in which case the holder would be able to exercise his patents to keep the goods out of other territories. In *Boussois*, the licensor did hold patents for the only process being used in seven Member States.

What most worried many practitioners were the reasons given by the Commission for stating that the patent licensing regulation did not apply to the agreement.[25] First, the know-how dominated the patented technology and did not merely permit a better exploitation of the licensed patents. This was a narrow interpretation of the regulation, since the qualification 'ancillary' had been deleted from the definition of a mixed know-how and patent licensing agreement in article 1. I shall argue at 5.1.4 below that the definition of 'ancillary' at article 10(15) of the technology transfer regulation may have changed the situation.

Secondly, the Commission said that the obligation to manufacture in the plant made for Boussois in France impliedly restricted manufacture in other Member States. The parties accepted that the licencee was confined to selling only in France although it needed no licence to produce elsewhere. The Commission treated this obligation as exceeding what is permitted by article 1 of the patent regulation, which exempts territorial protection only 'in so far and as long as' there is a valid patent in the territory to be protected.[26] Further territorial protection prevented the application of the exemption.

Nevertheless, the Commission granted an exemption as these provisions were necessary to persuade Interpane to grant a licence and Boussois to commit itself to the investment neededd to exploit the technology. The benefits made possible by the limitation of the licensee to a particular factory are spelled out in paragraph 20 of the decision and recital 10 of the technology transfer regulation rather better than in recitals 7 and 9 of the

---

[23] See 2.2–2.2.5 above. In para. 29 of the English text of *Boussois/Interpane* the Commission refers to 'the introduction of new technology' and the English text is the authentic one, but according to the German text it refers to the introduction of the original (*ursprunglich*) technology. I am indebted for this translation to Patricia Hollenstein. So it may be that *Nungesser* can be distinguished on the ground that this technology was not sufficiently new.   [24] See 2.1.2, n. 20, above and 7.3.14 below.
[25] See 1.2, text to nn. 35 and 36 above.
[26] Under the patent reg., there had to continue to be a valid patent in the licensee's territory for the agreement to come within the definition of a patent licensing agreement and one in the territory to be protected by the various restrictions permitted by art. 1(1). This is not the position under the know-how or technology licensing regulation under which the licence of know-how that continues to qualify as secret, substantial, and recorded may support the exemption.

know-how regulation. The market was competitive; the exclusive territory granted to Boussois, and the restrictions on its exploiting the technology elsewhere, made the parties more willing to negotiate a licence of important know-how. The territorial protection induced investment by Boussois to make, use, and put the product on the market, and so disseminate and further develop a new product, and increased the quantity and quality of the product produced.

Although the patent regulation did not apply, the Commission expressly referred to its recitals[27] in deciding that the territorial protection merited exemption. It no longer treated know-how as information to be communicated, but as a technology licence.

Consequently, the Commission found that the agreement contributed to technical progress and helped to disseminate a new product. Even the protection against passive sales by other licensees for five years, slightly longer than the period of five years from first putting on the market within the common market by licensor or licensee that were allowed by the patent regulation, was exempted since there were few Community suppliers, and well informed professional buyers would be able to seek the most favourable terms from the limited number of suppliers. Protection only from active sales would, therefore, not have sufficed to induce the necessary investments.

The Commission cleared some provisions: (1) the duty to keep the know-how confidential; (2) a non-exclusive and reciprocal duty to pass back improvements to the licensor; and (3) a duty to make payments in advance.

The decision was important for several reasons. In addition to the points already mentioned about the patent regulation not applying where the know-how dominates the patent, or where manufacture is not permitted in territories where there is no patent protection, it showed that the Commission was treating the communication of know-how as a technology licence, restrictions in which were justifiable on grounds similar to those in patent licences.

The decision in *Boussois/Interpane* that the payments in advance were outside the prohibition of article 85(1) was one of the earlier signs that the Commission has come to believe that royalties are a matter for the parties, in which it should not interfere.

Nevertheless the decision indicated that few patent licences would be

---

[27] In granting a group exemption, the patent reg. could not refer to the competitive nature of the market. The irrelevance of market power is one of the objections to group exemptions. In the USA, many lawyers and economists consider that vertical agreements should be *per se* legal in the absence of market power and capable of justification on grounds of efficiency even where the licensor enjoys such power. See 1.4.2 and 1.4.3 above.

exempt under the patent regulation, so it was necessary to proceed to prepare a group exemption that would cover pure know-how agreements and the mixed patent and know-how licences that were excluded.

3.1.4 *Rich Products/Jus-Rol*[28]

The Commission exempted a pure know-how licence, with the right to obtain improvements, concerning a process for making frozen yeast dough products. The licence provided for exclusive manufacture in the United Kingdom but all Rich Products' licensees in the common market were entitled to sell in any part of it. Each licensee was promised that Rich Products would not grant further production licences within its territory. One would have thought that this constituted an open exclusive licence within the meaning of the Court's judgment in *Nungesser*.[29] Transport costs, however, were substantial in relation to the value of the product, so parallel exports might not have been important.

The Commission found that the exclusive manufacturing licence infringed article 85(1). There were other processes for freezing dough, so the product was not new within the Court's ruling in *Nungesser*. This is difficult to reconcile with the Court's treatment of the INRA varieties of maize seed as new, overruling the view expressed in the Commission's decision that maize is old. Indeed, I have suggested that the Court did not limit its decision that open exclusive licences do not necessarily infringe article 85(1) to new technology or products. The INRA variety was new, but that was only one of the matters the Court thought important in ensuring inducements to invest in r & d.

'Newness' is a relative concept. Unless one accepts that know-how for which the licensee is prepared to pay must be new enough to qualify, great uncertainty is introduced into the law. It seems to me absurd for the Commission to say that, because there were substitute methods of preserving frozen yeast dough, the exclusive manufacturing licence infringed article 85(1) and required exemption. The other methods must have provided competition to the parties, and we are told in paragraph 4 that the rival processes were 'not easily accessible'. The Commission seems to have been reading the Court's clearance of 'open exclusivity' out of the law, although a recital in each of the three regulations accepts it.

In *Rich Products/Jus-Rol*, the Commission proceeded to exempt the exclusive licence on the ground that the exclusive manufacturing right was necessary to induce the licensees to invest—the ground on which, in *Nungesser*, an open exclusive licence was found not, in itself, to infringe

---

[28] [1988] OJ L69/21, [1988] 4 CMLR 527, CMR 10,956.
[29] Cited n. 1 and analysed at 2.2–2.2.5 above.

article 85(1). It may be, as I suggested at 2.2.5 and 3.1, that the Commission believed that it had to grant individual exemptions before it could adopt the regulation.

Nevertheless, the Commission cleared many provisions, such as the licensee's obligation not to use the know-how for ten years after termination, in so far as it has not entered the public domain. In some earlier decisions, it had adopted the theory that a restriction on using know-how after the term of a patent licence was not within the scope of the patent protection, and was, therefore, suspicious.[30]

The Commission cleared the licensee's obligation to use the know-how solely to make the licensed product, although the strength of the finding may be reduced, in that the Commission stated that the licensed product covered all possible uses of the know-how. Field of use restrictions have been whitelisted in all three regulations.[31]

The Commission also cleared the licensee's obligation to feed back improvements on a royalty free, non-exclusive basis for as long as the basic licence continued.[32]

The clearance of the licensee's obligation to obtain a secret pre-mix, necessary to ensure consistent quality, only from the licensor is very similar to the provision of the herbs in *Campari*.[33] The Commission does not treat the know-how contained in such a pre-mix as being licensed—it remains secret—so the know-how about the pre-mix would not qualify the licence as including substantial know-how under the regulation. Sufficiently important additional know-how probably was disclosed to Jus-Rol for the mark to be ancillary to the know-how, so the agreement might have qualified under the know-how regulation had it then been adopted.[34]

---

[30] In *Burroughs/Delplanque* and *Burroughs/Geha* [1972] OJ L13/50 and 53, [1972] CMLR D72, CMR 9485, analysed more fully in **Korah**, n. 1 above, however, the Commission cleared an 'obligation on the licensee not to use after the end of the agreement the technical know-how communicated to it during the period the agreement is in force'. It treated the exclusive provisions as *de minimis*, however, because licensor and both licensees could sell throughout the common market and together supplied only 10% of the market for carbon paper.

In an informal decision, *Cartoux/Terrapin*, *Tenth Report on Competition Policy*, point 129, the Commission stated that, following a complaint, it had persuaded the former licensor to amend such a ban on using the know-how after the term, so as to permit such use against the payment of reasonable fees for a reasonable period in respect of such know-how as remained secret. It did not state how reasonableness was to be assessed, or by whom.

A post-term use ban was originally blacklisted in the draft of the patent licensing reg., [1979] OJ C214/2, but was not listed in the final version, bringing such a ban under the opposition procedure if it restricts competition.

A restriction on using the know-how after termination is now whitelisted: art. 2(1) (3) of the know-how and technology transfer regs., discussed at 7.3.3 below.

[31] Art. 2(1)(8) of the technology transfer reg., 7.3.8 below.

[32] The proviso is reflected in art. 2(1)(4) of the technology transfer reg. and discussed at 7.3.4 below.　　　　　　　　　　　　　　[33] See 3.1.2 above.

[34] Contrast *Moosehead/Whitbread*, n. 18 above and described at 3.1.7 below, but see the position under the technology transfer reg. analysed at 5.1.4 below.

The Commission took a further step down the road of leaving the negotiation of royalties to the parties, discussed at 7.3.7 below. It cleared the licensee's obligation to pay royalties throughout the term, even if the know-how becomes known, since the licensee has the right to terminate after five years of the ten year term, and so free itself of the liability.

3.1.5 *Delta Chemie/DDD*

In *Delta Chemie/DDD*,[35] too, the sole and exclusive licence to manufacture in the United Kingdom, Ireland, and Greece and the limitation restraining manufacture elsewhere were exempted rather than cleared, although all licensees were permitted to sell throughout the common market. No reasons were given for finding that the agreement had the object or effect of restricting competition other than the weak ones given in *Boussois/Interpane* (3.1.3 above) and *Jus-Rol* (3.1.4 above). The Commission stated at paragraph 23 that this was not an agreement about new products, because they had already been distributed by the firm that later took a licence, some of them less than two years previously, others longer ago. Yet a distributor and its customers would not know the technology, even if they knew the product. The licence gave DDD access to the technology.

In *Nungesser*, the Court spoke both of products newly developed and unknown in the licensee's territory and of newly developed technology. I suggested at 2.2.3 that the novelty of the technology should be relevant when exclusivity is being justified by licensor's or licensee's investment in the original innovation or in preparing for production. It may be costly to start production by use of new technology, even if the product is already known, although it may be less expensive to develop a market.

The cost of developing a market has justified exclusive distribution agreements in individual decisions under article 85(3) and group exemptions,[36] but the Court accepted a justification under article 85(1) in relation to licences. Again the Commission may have exempted a sole and exclusive manufacturing licence rather than cleared it to establish its powers to adopt the group exemption, as I suggested at 3.1 above.

A new element in *Delta Chemie/DDD* was the arrangement by which the licensee, DDD, agreed to buy in the product from Delta Chemie while establishing its reputation and attracting sufficient demand to justify investment in production capacity. Thereafter, it was to be entitled to manufacture with Delta Chemie's know-how. When the decision was adopted DDD was already making one product, although it was still acting as a distributor for some of the other stain removers.

At paragraphs 38 to 40 the Commission found that regulation 1983/83,

---

[35] N. 8 above.   [36] In regs. 67/67 and 1983/83.

the group exemption for exclusive distribution agreements, did not apply even during the initial period during which DDD was acting as a distributor. The Commission considered that the distribution agreement was temporary and formed part of the know-how licence. Nevertheless it granted an individual exemption for almost twenty years, twice as long as is permitted under the block exemption for know-how agreements. Under regulation 1983/83, the exclusive territory and ban on active sales outside it could have lasted as long as the regulation with the expectation of automatic renewal.

There is nothing in the distribution block exemption to exclude an agreement limited in time. Nevertheless, the Court's ruling in *Pronuptia*,[37] that regulation 67/67 did not apply to distribution franchising agreements because franchising was different from distribution, might lead the Commission to treat as excluded from the regulation for exclusive distribution an agreement that it sees as ancillary to a know-how and trade mark licence. Most distribution agreements are closely connected with trade mark licences, and the Court ruled in *Hydrotherm*[38] that, provided the mark is not used to oppose parallel imports, the assignment of a trade mark to a distributor does not prevent the regulation from applying. It seems to have been the know-how licence that made the Commission believe that the exclusive distribution block exemption did not apply in *Delta Chemie/DDD*.

The know-how regulation expressly provided for the situation where the licensee starts developing a market by buying the product from the licensor, but the transfer of technology regulation no longer makes express provision for this situation. It provides in article 5(1)(5), as explained in recital 8, that territorial protection may not be granted under article 1(1) 'to agreements entered into solely for the purpose of sale'.[39] If *Delta Chemie/DDD* applies, the periods of protection provided by article 1(2) will start to run from the initial agreement and may expire before manufacture is started.[40] This may confer more territorial protection from passive sales than the Commission might have wanted on a dealer which changes its mind and never starts to produce itself. On the other hand, since the period permitted for protection from passive and, possibly, active sales may have expired during the dealing stage, the dealer may be dissuaded from starting to produce for lack of territorial protection when it does so, thereby reducing the incentive for the dealer to start production.

---

[37] *Pronuptia de Paris GmbH* v. *Pronuptia de Paris Irmgard Schillgalis* (161/84), [1986] ECR 353, [1986] 1 CMLR 414, CMR 14,245, para. 33.

[38] *Hydrotherm Gerätebau GmbH* v. *Compact de Dott. Ing. Mario Andreoli & C Sas* (170/83), [1984] ECR 2999, [1985] 3 CMLR 224, CMR 14,112, paras. 17–22.

[39] Analysed at 5.1.2 below.   [40] Analysed at 6.6.5–6.6.7 below.

There is no mention of a proposal to manufacture in the patent licensing regulation which was relevant when *Delta Chemie* was decided, but common sense dictates that the licensee must be given time to establish its production facilities. During that time it may, presumably, enjoy the territorial protection against active and passive sales permitted by the technology transfer regulation.

If it is envisaged that the licensee will start by buying in the product until it has developed a market and set up production, can the permissible territorial protection be prolonged by keeping separate the distribution and licensing agreements?[41] One might, initially, grant the distributor an exclusive sales territory under regulation 1983/83, giving it an option to take a know-how licence should it later decide to manufacture. It would be necessary to make the distribution agreement independent of the licence as, otherwise, the licence would be treated as part of the original distribution agreement[42] and the precedent of *Delta Chemie/DDD* might apply.

Alternatively, if dealers need protection from passive sales one might consider drafting the standard agreement as a know-how licence, but arranging for the dealer to buy from the licensor for an initial period. Where this is clearly a sham, a court or the Commission might hold that the agreement is not a know-how licensing agreement, but there may be cases where the possibility of subsequent manufacture is substantial.

3.1.6 *Transocean Marine Paint*

A series of exemptions has been granted to the *Transocean Marine Paint Association*.[43] Since subsequent coats of paint will adhere satisfactorily

---

[41] Sebastiano Guttuso, speaking in his personal capacity at a conference organized in London by the European Business Institute on 20 Mar. 1989, suggested that this may be possible.

[42] See, e.g., *BP Kemi/DDSF* [1979] OJ L286/32, [1979] 3 CMLR 684, CMR 10,165, para. 46, where the Commission treated the co-operation and purchasing agreement, which came into effect on the same date, together with a later addendum, as a single agreement because the co-operation agreement referred to the purchasing agreement as the main contract, and the provisions of the co-operation agreement were a necessary counterpart of the exclusive purchasing obligation. It seems that neither would have been signed without the other.

The fact that the licence is dependent on the option in the distribution agreement is probably not enough for the agreements to be treated as a single agreement as long as there is no obligation, enforceable or otherwise, on the licensee as well as on the licensor, to enter into the subsequent licensing agreement.

In *Moosehead/Whitbread*, cited at n. 18 above and described at 3.1.7 below, the three contracts that were treated as a single agreement came into operation at the same time and in *Delta Chemie/DDD*, the preliminary distribution agreement was made at the same time as the know-how agreement.

[43] The last decision was reported in [1988] OJ L351/40, [1989] 4 CMLR 621, [1989] 1 CEC 2003. The three earlier decisions are cited in the recitals to the 4th.

only to paint of a similar chemical composition it is important that a firm making or selling marine paint should be able to provide paint which a ship's master can pick up in any port the ship is likely to visit. Consequently, the association, which has members in many parts of the world, communicates to each of its members the recipe for making paint and licenses them to use its mark. Each member receives an exclusive territory and was originally required to pass over to a member in another territory where it sold the paint some of the profit earned on sales made to customers there. This profit passover is no longer required, but an active sales policy outside each territory is restrained.

The Commission exempted the agreement for a fourth time on the ground that the limited exclusivity obliged members to concentrate their efforts within the allotted territory, and was necessary to make the Transocean network viable. On grounds of confidentiality the market share of the association has been deleted from the decision as published, but members meet competition from larger and more powerful suppliers of similar products.

This may be an example of a know-how pool between actual or potential competitors, with ancillary trade mark rights, where some territorial protection against active sales was exempted. It would be excluded from the automatic application of all three group exemptions by article 5(1)(1) to (2) if the transaction be treated as a patent pool or a joint venture. It might also have been excluded from the know-how regulation on the ground that the know-how was ancillary to the trade mark licence and not the other way round, but this may well have been changed in the technology transfer regulation by the unusual definition of 'ancillary' in article 10(15). See 5.1.4 below.

3.1.7 *Moosehead/Whitbread*[44]

Moosehead made beer in Canada with a special yeast and sold it under the mark 'Moosehead'. It is difficult for the supplier of a brand not known in England to enter the market because most beer is sold in draft mainly in pubs. The English licensing laws create an important barrier to the creation of new pubs and most pubs are tied to one or other of the brewers. Moreover, most brewers distribute directly to the pubs, discouraging the development of wholesalers. Consequently, foreign brewers sometimes enter into a trade mark licence with a local brewer which has a substantial network of tied pubs.

Moosehead granted Whitbread the sole and exclusive right to produce and market beer under the Moosehead mark in the British Isles. It also

---

[44] N. 18 above.

transferred its mark for the United Kingdom into their joint names, and provided Whitbread with the specific yeast and recipe. Whitbread agreed that the type and quality of the beer produced under the contract would comply with the Moosehead specifications and not actively to seek customers for the product outside its territory, although it was free to accept unsolicited orders. It also agreed, during the term of the agreement, not to produce or promote within the territory any other beer identified as a Canadian beer.

Moosehead agreed to provide Whitbread with the special yeast and all the relevant know-how for producing the beer, while Whitbread agreed to comply with Moosehead's directions, to buy the yeast only from Moosehead or a person designated by Moosehead, to use the know-how only for the production of the product, and to keep it confidential. Lawyers might call this 'an exclusive know-how and trademark licence' and their clients 'an industrial franchise'.

The Commission's realistic analysis under article 85(1) in relation to the promise by Whitbread not to challenge Moosehead's right to the mark and as to Whitbread's obligation not to contest its validity was very welcome. It said:

15.4. (a) . . . A clause in an exclusive trade mark licence agreement obliging the licensee not to challenge the ownership of a trade mark, as specified in the above paragraph, does not constitute a restriction of competition within the meaning of Article 85(1). Whether or not the licensor or licensee has the ownership of the mark, the use of it by any other party is prevented in any event, and competition would not thus be affected.

The validity of a trade mark may be contested on any ground under national law, and in particular on the grounds that it is generic or descriptive in nature. In such an event, should the challenge be upheld, the trade mark may fall within the public domain and may thereafter be used without restriction by the licensee and any other party.

Such a clause may constitute a restriction of competition within the meaning of article 85(1), because it may contribute to the maintenance of a trade mark that would be an unjustified barrier to entry into a given market.

Moreover in order for any restriction of competition to fall under article 85(1), it must be appreciable. The ownership of a trade mark only gives the holder the exclusive right to sell products under that name. Other parties are free to sell the product in question under a different trade mark or trade name. Only where the use of a well-known trade mark would be an important advantage to any company entering or competing in any given market and the absence of which therefore constitutes a significant barrier to entry, would this clause which impedes the licensee to challenge the validity of the trade mark, constitute an appreciable restriction of competition within the meaning of article 85(1).

(b) In the present case Whitbread is unable to challenge both the ownership and the validity of the trade mark.

As far as the validity of the trade mark is concerned it must be noted that the trade mark is comparatively new to the lager market in the territory. The maintenance of the 'Moosehead' trade mark will thus not constitute an appreciable barrier to entry for any other company entering or competing in the beer market in the UK. Accordingly, the Commission considers that the trade mark no-challenge clause included in the agreement, in so far as it concerns its validity . . . does not constitute an appreciable restriction of competition and.does not fall under Article 85(1).

Furthermore, in so far as this clause concerns ownership, it does not constitute a restriction of competition within the meaning of Article 85(1) for the reasons stated in the first indent of point 15.4 above.

The exclusive trade mark licence with a restriction on actively selling outside the territory and a restriction on competing with the Moosehead mark within the territory was, however, found to restrict competition contrary to article 85(1). The exclusive licence foreclosed the other five big brewers from obtaining a licence from Moosehead. The effect on trade between Member States and on competition was appreciable, as Whitbread was large enough to export to other Member States.

This part of the legal appraisal is difficult to reconcile with the Commission's view about the restraint on challenging the validity of the mark and seems to be a return to the *per se* approach to exclusive territories adopted in earlier decisions. The mark was new in the territory and the Commission said that it could not have created much of an entry barrier for the other big brewers.

Equally important for technology transfer, the Commission found that the licence did not qualify under the group exemption for know-how licences:

(16) 1. The block exemption provided by Commission Regulation (EEC) No. 556/89 applies to agreements combining know-how and trademark licences where, as stated in Article 1(1), the trademark licence is ancillary to that of the know-how. In the present case the principal interest of the parties lies in the exploitation of the trademark rather than of the know-how. The parties view the Canadian origin of the mark as crucial to the success of the marketing campaign, which promotes the Product as Canadian beer. Under these circumstances, the provision of the agreement relating to the trademarks is not ancillary and Regulation 556/89 therefore does not apply.

This view has prevented industrial franchises qualifying under the know-how regulation even where there was sufficient technical know-how to qualify as 'substantial'. The Commission's position was extreme, because the mark was new in Europe, so the licence was no more crucial than is the mark in any franchising agreement. Usually, the know-how and the mark are complementary to each other. Neither is much use without the other. In such cases, the Commission has held that the group exemption does not apply.

It may be argued that industrial franchises may now come within the ambit of the technology transfer regulation because of the definition of 'ancillary provision' in article 10(15). This is, however, controversial since recital 6 speaks of extending the regulation to intellectual property rights other than patents 'when such additional licensing contributes to the achievement of the objects of the licensed technology and contains only ancillary provisions'. See 5.1.4 below.

The Commission was considering preparing a separate group exemption for franchise agreements after the adoption of the transfer of technology regulation. This may now be unnecessary.

3.1.8 A Possible Change in the Commission's View

Officials in the Commission disagree strongly about whether it should be concerned by vertical agreements when the market is competitive and insufficiently concentrated for collusion to be possible. Some say that in those circumstances vertical agreements cannot be anti-competitive: others that vertical exclusive agreements have contributed significantly to the isolation of national markets, because so often the territories correspond with the boundaries of Member States. The Commission is currently trying to investigate the effects of its strong policy against territorial restraints imposed vertically and is expected to publish a Green Paper after the summer of 1996.

I hope that the Commission's policy may change. It still lacks adequate resources to vet all exclusive agreements, and many officials now accept the Court's view about ancillary restraints being outside article 85(1).[45] Provided that the officials dealing with individual transactions that may increase competition are prepared to draft comfort letters stating that the agreement does not restrict competition contrary to article 85(1) it will be easier to enforce in national courts ancillary restraints that make a transaction viable. If such a policy were well publicized, this could also lead to a dramatic reduction in the number of agreements notified.

3.2 LIMITED LICENCES

3.2.1 Limited Patent Licences

In 1962,[46] the Commission took the view that limited patent licences did

---

[45] See 1.4.1 and 2.2–2.2.5 above. The Commission now accepts that the ancillary-restraint doctrine applies to joint ventures, although the Court has not had occasion to apply the doctrine in that field. See the Commission's notice concerning the assessment of co-operative joint ventures pursuant to art. 85 of the EEC Treaty [1993] OJ C43/2, [1996] 4 CMLR 357.

[46] See notice on patent licensing agreements, JO 2922/62, commonly known as 'the Christmas Message', partly because of its date and partly because of the good news that many

not infringe article 85(1). It was the patent that restrained the licensee from exceeding the limitations—the licence merely permitted it to do what would otherwise have been unlawful:

> I. On the basis of the facts known at present, the Commission considers that the following clauses in patent licensing agreements are not caught by the prohibition laid down in article 85(1) of the Treaty:
>
> A. Obligations imposed on the licensee which have as their object:
>
>> 1. Limitation of the exploitation of the invention to certain of the forms which are provided for by patent law (manufacture, use, sale).
>> 2. Limitation:
>>> (a) of the manufacture of the patented product,
>>> (b) of the use of the patented process, to certain technical applications;
>> 3. Limitation of the quantity of products to be manufactured or of the number of acts constituting exploitation.
>> 4. Limitation of exploitation:
>>> (a) in time
>>> (a licence of shorter duration than the patent);
>>> (b) in space
>>> (a regional licence for part of the territory for which the patent is granted, or a licence limited to one place of exploitation or to a specific factory);
>>> (c) with regard to the person
>>> (limitation of the licensee's power of disposal, e.g., prohibiting him from assigning the licence or from granting sub-licences)
>
> . . . .
>
> IV The obligations listed at I(A) do not fall under the prohibition of Article 85(1) because they are covered by the patent. They entail only the partial maintenance of the right of prohibition contained in the patentee's exclusive right in relation to the licensee, who in other respects is authorized to exploit the invention. The list at 1(A) is not an exhaustive definition of the rights conferred by the patent.

By the time of the first decisions on technology licences in 1972 the Commission came to see patent licences as horizontal, at least when both licensee and licensor have started to exploit the invention.[47] It began to question the German theory of the inherent right of the patent which underlay the Christmas Message. In none of its formal decisions on patent

exclusive licences would not infringe art. 85(1) and need not be notified to the Commission under reg. 17. It was withdrawn when the group exemption for patent licences was adopted.

[47] See 1.4.2 above.

licences did the Commission accept the view it had announced in 1962 that limited licences were not anti-competitive save on *de minimis* grounds.

In particular, the Commission denied that territorial limitations escaped article 85(1). The Court had ruled in *Centrafarm* v. *Sterling*[48] that a patentee could not use its Dutch patent to exclude products sold by it or with the holder's con sent in the United Kingdom.

The language of the Court's ruling and opinion referred to marketing in the territory of export by or with the consent of the holder, although where a licensee sold directly into the country where the intellectual property right was being enforced, the holder would have had a chance to earn a reward—part of the specific subject matter of a patent.

The Commission interpreted this to mean that the grant of a licence exhausted a patent: a licence to make in one factory or in one Member State amounted to a licence to make and sell throughout the Community subject to contractual territorial restrictions that infringed article 85(1).[49] This went further than the case law of the Court in that it is unclear whether the grant of a patent licence exhausts the rights, or only a sale by the licensor or with his consent does so.[50] In the Commission's view, the

---

[48] *Centrafarm BV* v. *Sterling Drug Inc.* (15/74), [1974] ECR 1147, [1974] 2 CMLR 480, CMR 8246. Discussed at 2.1.2.

[49] e.g., in its *Fifteenth Report on Competition Policy* (Brussels: EC Commission, 1986), point 81, at 82, the Commission said:
> 'In the Commission's opinion, a licence granted by a Community copyright protection society [limited to France] is valid throughout the Community, and authorizes manufacture, even by way of custom pressing, in any Member State.'

This repeated the view it had stated in a press release relating to *GEMA*, *Fifteenth Report on Competition Policy*, n. 49 above, point 81.

[50] In *Musik-Vertrieb Membran* v. *GEMA* [1982] 1 CMLR 630, the German Supreme Court, the Bundesgerichtshof, applying EEC law, on the basis of the Court's ruling in the same case (55 and 57/80), [1981] ECR 147, [1981] 2 CMLR 50, held that direct sales by a copyright licensee from its territory into West Germany could be restrained by use of the German copyright. Where the patentee granted a licence to manufacture and sell in one country, or under the patent in that country, it had not consented to the sale elsewhere.

The matter was raised in *Pharmon* v. *Hoechst* (19/84), [1985] ECR 2281, [1985] 3 CMLR 755, CMR 14,206; comment **Guy J. Pevtchin** and **Leslie Williams**, 'Pharmon v. Hoechst: Limits on the Community Exhaustion Principle in respect of Compulsory Licences', in **Hawk** (ed.) *Thirteenth Annual Proceedings of the Fordham Corporate Law Institute* (New York: 1986), ch. 12. Mancini AG at 2285–6 advised the Court to rule that, where goods were made without the consent of the patentee under a compulsory licence and sold directly by the licencee in another Member State, the rules for the free movement of goods do not prevent the exercise of patent rights to restrain their import. The Court's judgment is clearly limited to compulsory licences and does not establish the effect of the rules for free movement on direct sales by voluntary licensees outside their territories. At 2285–6 the AG, however, considered that where the holder has consented to a licence, it can no longer oppose even if the licensee sells directly into the country where the patent rights are being exercised. The question remains open.

The Commission may well now consider that the grant of a patent licence does not exhaust. See art. 2(1)(14) of the technology transfer reg., analysed at 7.3.14 below.

patent did not exclude parallel imports because the grant of a licence exhausted the intellectual property right.

In many decisions, the Commission said that licences limited territorially were to be treated as contractual restrictions on exceeding the limit and were subject to article 85(1).[51] Nevertheless, in all three regulations[52] 'exploitation' was defined disjunctively, in terms very similar to article 10(10) of the technology transfer regulation.[53] This seems to be an example[54] of the Commission accepting the theory of a limited licence. Under article 1(1)(8) the holder may license X to manufacture for its own purposes, retaining the exclusive right to sell within a territory. This might, but need not, be licensed to someone else.

---

[51] In *AOIP* v. *Beyrard* [1976] OJ L6/8, [1976] 1 CMLR D14, 22, CMR 9801, the Commission treated a patent and know-how licence to make in France together with permission to export to any country where the patentee had not granted a licence as a contractual ban on exporting. It repeated this view in *Re the Agreements between BBC Brown Boveri and NGK Insulators Ltd* [1988] OJ L301/68, [1989] 4 CMLR 610, para. 18: a licence to manufacture and sell in Japan and the Far East implied a ban on exporting to the Member States where BBC held no intellectual property rights. In a press release relating to *GEMA*, n. 49 above, the Commission stated that:

> 'a license granted by a Community copyright protection society is valid throughout the Community, and authorizes manufacture, even by way of custom pressing, in any Member State.'

See also *Boussois/Interpane*, n. 20 and described at 3.1.3 above, where a licence to manufacture in facilities provided by the licensor in France was treated by the parties and the Commission as a restriction on manufacturing elsewhere in the common market; and in *Delta Chemie/DDD*, n. 8 above and described at 3.1.5 above, at para. 25, the Commission stated that an exclusive manufacturing right implies, 'by its very nature', that it [the licensee] is prohibited from manufacturing these products outside the licensed territory. See also *Rich Products/Jus-rol*, n. 28 and described at 3.1.4 above. In *Windsurfing* [1983] OJ L229/1, [1984] 1 CMLR 1, para. 98, the Commission treated a patent licence to manufacture at a specific address in West Germany as an obligation not to manufacture in other Member States, even when there was no patent protection and no need for a licence. On appeal, (193/83), [1986] ECR 611, [1986] 3 CMLR 489, CMR 14,271, from para. 82, the Court's judgment is not very clear, but may be construed to mean that such a licence should be treated as a contractual restriction on manufacture elsewhere.

In *Nungesser*, n. 1 above, para. 53, discussed at 2.2–2.2.5 above, the Court was inconsistent about whether restricting licensees from selling in each other's territories amounted to open exclusive licences or conferred absolute territorial protection.

The Commission's formalistic attitude to territorial restrictions and, in particular, its distinction between the lesser protection allowed from licensees than from the licensor, was criticized by **René Joliet**, shortly before he became a member of the Community Court, in 'Territorial and Exclusive Trade-Mark Licensing under the EEC Law of Competition' (1984) 15 *IIC* 21, 35. The licensee may need protection from all directions, and not merely from one. A bucket in which four holes have been mended will still not hold water if a fifth hole remains.

[52] '(2) patent licensing agreements are agreements whereby one undertaking, the holder of a patent (the licensor) permits another undertaking (the licensee) to exploit the patented invention by one or more of the means of exploitation afforded by patent law, in particular manufacture, use or putting on the market.'

[53] Set out at 3.2.4 below.

[54] For other examples see 3.2.5, nn. 63 and 64 below.

3.2.2 Limited Trade Mark Licences

Following the judgment of the Court in *Hag I*[55] the Commission has treated trade marks as being exhausted by the grant of a licence.[56] *Hag I* was expressly overruled by the Community Court in *Hag II*[57] on other grounds, and it is not clear how far the precedent of *Hag I* remains useful. In *Ideal-Standard*[58] the Community Court seemed to assume that a trade mark would be exhausted by a licence of the mark for one country, but held that it was not in the instant case where the mark had been assigned together with the sale of a business and the assignor retained no control over the products to which the mark was attached because of the risk of confusing consumers. If direct import could not be restrained consumers might be confused about the origin of the products.

It is not clear how far these cases on trade marks, where the ability to control the products to which the mark is applied in order to prevent confusion is vital, affect the case law on the exhaustion of patent rights.

3.2.3 Limited Know-how Licences

Some say that, in theory, it is impossible to apply the concept of limited licences to know-how. There is no exclusive right equivalent to the inherent right of the patent. Once the technology has been communicated, any restriction on exploitation must be by contract and subject to article 85. Once the technology has been communicated to the licensee, it can use it as it will, subject only to contractual restrictions.

Others, of a less formalistic turn of mind, say that the theory of limited licences does apply to know-how. The licensee's ability to use the know-how derives from the contract without which it would not have learned the technology.[59] If the contract limits those rights in any way, it does not permit use outside the limits.

Moreover, business has persuaded the Commission to think in terms of

---

[55] *Van Zuylen Frères* v. *Hag AG* (192/73), [1974] ECR 731, [1974] 2 CMLR 127, CMR 8230.

[56] *Velcro/Aplix* [1985] OJ L233/22, [1989] 4 CMLR 157, CMR 10,715.

[57] *CNL Sucal* v. *Hag GF AG* (C–10/89), [1990] ECR I–3711, [1990] 3 CMLR 571, [1991] 2 CEC 457.

[58] *IHT Internazionale Heiztechnik GmbH* v. *Ideal-Standard GmbH* (C–9/93), [1994] ECR I–2789, [1994] 3 CMLR 857, [1994] 2 CEC 222.

[59] Some competition lawyers object on the grounds that the licensee might have developed the technology independently or taken a licence from someone else if it had not accepted a limited licence. Others would say that the point is relevant only if the parties were potential competitors before the grant of the licence. Whether the licensee could have developed or acquired competing technology independently may be difficult to ascertain. Moreover, even if the licensee could have done so, the technology licensed may have saved it significant time and money in getting its production going. It will have given it a head start.

licensing technology and not to make formalistic distinctions based on whether or not there is patent protection. The dissemination and exploitation of unprotected know-how also requires investment, which might not be undertaken without protection from free riders. Indeed, the very fragility of know-how—there is no exclusive right to rely upon once it reaches the public domain—requires that greater freedom of contract be allowed. There are several signs in the know-how and technology transfer regulations that the Commission is accepting the limited licence theory, apart from territorial limitations, which it believes might delay the integration of the market. See 3.2.5 below.

3.2.4 'Exploitation' Defined Disjunctively (Article 10(10))

'Exploitation' has been defined in terms similar to the patent licensing regulation:

> (10) 'exploitation' refers to any use of the licensed technology in particular in the production, active or passive sales in a territory even if not coupled with manufacture in the territory, or leasing of the licensed products;

Since all three regulations define 'exploitation' disjunctively, one may infer that the holder of technology may license any one of the methods of exploitation to different licensees. This permits licences limited to production or, probably, to active or passive sales, or to leasing to be granted to different undertakings, although, unless the licensee proposes to manufacture, much of the territorial protection exempted by article 1(1) is inapplicable.[60] It would, however, be pointless for the regulation to limit the territorial restrictions permitted by article 1(1) in complex ways, if this could be circumvented by granting licences limited territorially. Article 10(10) probably does not provide for this.

Where there are exclusive intellectual property rights, the holder may grant a licence to manufacture only, reserving the right to sell. The manufacturing licensee may use the product as a raw material or component of something it makes itself, in which case the regulation may be useful. Alternatively, the manufacturer may agree to deliver only to the licensor, although such a transaction might well be covered by the sub-contracting notice[61] and need no exemption. It is arguable that the holder can also appoint the sub-contractor as its exclusive distributor for a territory under regulation 1983/83 without actually taking delivery, so the sub-contractor can agree not actively to sell outside its territory for the

---

[60] See 5.1.2–5.1.3 below and *Delta Chemie/DDD*, cited at n. 8 and described at 3.1.5 above.

[61] [1979] OJ C1/2.

duration of the agreement.[62] A similar transaction is possible in the case of know-how, where there is no exclusive property right, because of the disjunctive wording of article 10(10).

3.2.5 Conclusion on Limited Licences

The Commission no longer assumes that any conduct that would not amount to infringement of an intellectual property right can be restrained only by contract and is automatically contrary to article 85(1).

There are some examples of the Commission's acceptance of the limited-licence theory in its decisions[63] and the regulation.[64]

The most significant change in the Commission's thinking brought to light by the latest regulation is contained in article 2(1)(14), analysed at 7.3.14 below. I shall argue that it is now asserting that the grant of a licence does not exhaust even an intellectual property right. This corresponds to the language of the Court's rulings on free movement.

---

[62] I see no reason why these two arrangements being made at the same time and probably constituting a single agreement should prevent both the sub-contracting notice and reg. 1983/83 from applying.

In *BP/Kemi/DDSF*, n. 42 above, the Commission held that two contracts, signed on the same day and made between the same firms, were part of a single agreement, since one was the counterpart of the other—neither would have made commercial sense without the other. That would not necessarily be the case in the example given in the text. It might make sense for the innovator to have the goods made by a sub-contractor to which it gave the necessary technology. The decision to use the same firm as distributor might well be a separate decision. Whether both were part of a single agreement would depend on the facts.

[63] Even after the Commission ceased to accept what it had said in its Christmas Message, an obligation of confidentiality was always cleared. More recently, see, e.g., *Delta Chemie/DDD*, n. 8 above, para. 31, where a field-of-use restrictions for the period when the technology was still secret was cleared on the ground that

'it constitutes the corollary to the acknowledged right of the licensor to dispose freely of his know-how and, as a result, to limit the use by third parties solely to the manufacture of the licensed products. If this were not the case the holder of the right may be deprived of a more or less important part of the receipts from his know-how.'

The Commission also cleared a post-term use ban which limited the licence in time, a duty of confidentiality, and a restriction on sub-licensing. Similar clauses were also cleared in *Rich Products/Jus-Rol*, cited at n. 28 and described at 3.1.4 above. See also *Mitchell Cotts/Sofiltra*, cited at n. 4 and described at 3.1.1 above.

[64] e.g. art. 1(1)(8) exempts, but does not clear, a restriction on production for one's own use only.

— Art. 2(1)(1) and (2) whitelists a duty of confidentiality and restrictions on sub-licensing;

— a ban on using the technology after the licence has come to an end is whitelisted by art. 2(1)(3) on the ground, stated in recital 20, that one must be able to lease as well as sell one's know-how;

— field-of-use restrictions are whitelisted by art. 2(1)(8), at least for different technical fields of use and different roducts;

— a restriction on using the know-how to construct plant for third parties is whitelisted by art. 2(1)(12);

3.3 HARMONIZATION OF INTELLECTUAL PROPERTY RIGHTS[65]

The Court's extension of the doctrine of exhaustion of intellectual property rights did much to prevent their exercise dividing the common market,[66] but where rights subsist in one Member State and not in another, the holder could exercise his rights to prevent the import of goods sold without his consent by third parties in countries where he had no right. Nancy Keen was able to keep out the goods that had been put on the market in Germany, where it had no design right and had not consented to the sale.[67] Warner was able to exercise its rental right over the cassettes in Denmark, although they had been legally acquired in the United Kingdom before a rental right was introduced.[68] When copyright lasted for the life of the author plus seventy years in Germany, but only for life plus fifty years elsewhere, the copyright material could be kept out of Germany for the final twenty years.[69]

To minimize the division of the common market in this way, the Commission proposed and the Council[70] has adopted several directives harmonizing national intellectual property laws. Usually the protection that Member States are required to give is at the level of that in the Member State granting most protection. So, copyright must now last for life plus seventy years.

The process of harmonization has been extremely political. The Commission has harmonized at the level of greatest protection, and lobbying on both sides by industry has been strenuous.

This monograph, however, is concerned with technology licensing and will not consider harmonization in detail. It may be worth mentioning,

---

 — quantity restrictions are permitted when a licence is granted to provide a second source by virtue of art. 1(2)(13).
 — art. 1(1)(14) assumes that a licence to sell in one area does not imply a licence to sell elsewhere;
 — 'exploitation' in all three regulations has been defined disjunctively. See 3.2.4 above.

[65] On harmonization of copyright generally, see **Hugh Laddie**, **Peter Prescott**, and **Mary Vitoria**, *The Modern Law of Copyright and Designs* (2nd edn., London: Butterworths, 1995).
 The harmonizing legislation is set out in convenient form in **Anna Booy** and **Audrey Horton** (eds.), *EC Intellectual Property Materials* (London: Sweet & Maxwell, 1994).
[66] See 2.1.2–2.1.3 above.
[67] *Keurkoop BV* v. *Nancy Kean Gifts BV* (144/81), [1982] ECR 2853, [1983] 2 CMLR 47, CMR 8861.
[68] *Warner Bros. and Metronome* v. *Christiansen* (158/86), [1988] ECR 2605, [1990] 3 CMLR 684, [1990] 1 CEC 33.
[69] *EMI Electrola* v. *Patricia* (341/87), [1989] ECR 79, [1989] 2 CMLR 413, [1990] 1 CEC 322.
[70] Since the Maastricht Treaty, the co-operation of the European Parliament has also been required under art. 189(b).

however, that the Commission is now interpreting unclear provisions in the various directives to require Member States to abrogate international exhaustion under national law where the products were first put on the market outside the Common market or EEA.

For instance, the Commission treats article 7 of the first trade mark directive[71] as requiring Member States not to adopt a doctrine of international exhaustion under national law. Consequently, Member States are amending their laws so that the holder will be able to exercise intellectual property rights to restrain the import into the common market of products put on the market outside the Community by him or with his consent. It is thought that the Commission is taking this line in order to enable holders to market goods or license technology to Eastern Europe, where only low prices are obtainable, without undermining their market in the common market.

3.4 THE THREE GROUP EXEMPTIONS FOR TECHNOLOGY LICENCES

The group exemption for patent licensing expired at the end of 1994 and was twice renewed until the end of 1995 while the Commission completed its consultations on the new regulation.[72] The Commission decided to simplify the regulatory structure and adopt a single regulation to apply to pure patent agreements, pure know-how agreements, and mixed technology

---

[71] First Council dir. 89/104 of 21 Dec. 1988 to approximate the laws of the Member States relating to trade marks: [1989] OJ L40/1.

'1. The [national] trade mark shall not entitle the proprietor to prohibit its use in relation to goods which have been put on the market in the Community under that trade mark by the proprietor or with his consent.'

Germany and Belgium have altered their national law to abrogate international exhaustion, but the Scandinavians are taking the line that it is not required.

The matter has been referred to the Community Court in *Phyterion* v. *Bourdon* (C–352/95). There the product was acquired in Turkey and imported into Germany. It was then sold and exported into France. The French trade mark holder objected because the product had never been put on the market in the European Community with its consent. The holder argues that the principle of international exhaustion under French law is contrary to art. 7 of the trade mark dir.

[72] It was renewed retrospectively for 6 months in Jan., 2 weeks after its expiry and again in Sept. over 2 months after its second expiry. Art. 11(2) of the transfer of technology regulation extended it again until the end of Mar. 1996 and art. 11(3) provides that:

'The prohibition in article 85(1) of the treaty shall not apply to agreements in force on 31 March 1996 which fulfil the exemption requirements laid down by regulations (EEC) No 2349/84 or 556/89.'

Now that the ceilings on the parties' market shares no longer prevent the application of the exemption, the sloppy prolongation of the patent reg. may not be disastrous, as almost any agreement drafted to come within it will come within the new reg. as from Apr. 1996. Problems could arise where proceedings were commenced during the periods when the regulation had expired and not yet been renewed retrospectively. See 9.2.1 below.

licences including both. This saves both firms and the Commission from having to decide whether the know-how was ancillary to the patent, in which case the patent regulation could apply, and from making careful assessments of which regulation was more favourable when the economic concerns were similar.

Transitional provisions will prolong until the end of 1999 the exemption of agreements that qualified under either of the earlier regulations before 1996. I have decided not to explain all the provisions applying to such agreements, as they are analysed in my earlier monographs on the two regulations, published by ESC Publishing.[73] The changes will, however, be picked up.

3.4.1 *Vires*

All three regulations were made under the same powers as the group exemptions for exclusive distribution and purchasing. Article 1(1) of regulation 19/65 enables the Commission to declare article 85(1) of the Treaty inapplicable to:

> categories of agreements to which only two undertakings are party and: . . .
>
> (b) which include restrictions imposed in relation to the acquisition or use of industrial property rights—in particular patents, utility models, designs or trade marks—or to the rights arising out of contracts for assignment of, or the right to use, a method of manufacture or knowledge relating to the use or to the application of industrial processes.[74]
>
> (2) The regulation shall define the categories of agreements to which it applies and shall specify in particular:
>
> (a) the restrictions or clauses which must not be contained in the agreements;
>
> (b) the clauses which must be contained in the agreements, or the other conditions which must be satisfied.

Recital 4 requires that, before adopting such a regulation, the Commission must gain experience through adopting individual decisions. At 3.1, I suggested that this may explain why exclusive territories in all the know-how licences have been exempted by the Commission rather than cleared, even though some of the agreements seem to grant open exclusive licences within the meaning of the Court's judgment in *Nungesser*.

---

[73] N. 1 above; *Know-how Licensing and the EEC Competition Rules—Regulation 556/89* (Oxford: ESC Publishing, 1989). ESC has since been acquired by Sweet & Maxwell. Both works are out of print, but available in some liraries.

[74] This phrase is wide enough to cover know-how that is not protected by intellectual property rights, so there is no doubt about the validity of the reg. as a whole on the basis of its *vires*.

When enforcing an agreement, one may argue that technology licences, however, are not required to come within the terms of this regulation.[75] They may escape the prohibition of article 85(1) because all the restrictive terms are ancillary restrictions necessary to make viable a pro-competitive transaction, or because even the network of agreements does not affect trade between Member States. They may be the subject of an individual decision exempting them.

3.4.2 The Structure of the Three Regulations

The Commission decided in 1994 to grant a single group exemption for pure patent licences, pure know-how, and mixed patent and know-how licences. The regulation is far more intricately drafted than the US law.[76] It follows closely the structure of the previous two regulations that exempted patent and know-how licences, although some of the recitals have been omitted or changed.

Article 1(1) exempts exclusive territories, ancillary export restraints, and two other obligations. The regulation applies only if one of these obligations, or a similar provision of more limited scope, is included. The main change originally proposed in the draft technology transfer regulation was the ceilings imposed on the market shares of the licensee by article 1(5) and (6). After a furore and effective lobbying by industry, these have now been abrogated. Consequently, few agreements that came within the earlier regulations will fail to come within the new one.

The duration of the territorial restraints where the licence includes know-how will often be longer than under the earlier regulations. The permitted periods start only from the date when the products were first put on the market by a licensee.

Article 2 contains the white list of provisions that are stated rarely to infringe article 85(1), but which are exempted just in case they may do so in particular circumstances. This creates greater certainty, and the governing recital may be helpful when considering licences that cannot be brought within the regulation, such as those of marks, copyright or software where these licences are not ancillary to the patent or know-how licence. The white list has been lengthened in the latest regulation.

---

[75] In *VAG France SA* v. *Etablissements Magne SA* (10/86), [1986] ECR 4071, [1988] 4 CMLR 98, CMR 14,390, the Court ruled that exclusive dealing agreements for motor vehicles were not required to come within reg. 123/85. Bringing an agreement within the terms of a group exemption is merely one way of avoiding the prohibition of art. 85(1).

It repeated this view in *Grand Garage Albigeois SA and Others* v. *Garage Massol SARL* (C–226/94), judgment of 15 Feb. 1996.

[76] 1.4.3 above.

Article 3 constitutes the black list of provisions or circumstances when the regulation will not apply. This has been shortened in the technology transfer regulation.

Article 4 provides the opposition procedure. If a licensing agreement within the meaning of article 1(1) contains a provision restrictive of competition which is not listed in any of the first three articles, the parties may notify the Commission, and if it does not oppose the application of the regulation within six months (four months under the latest regulation) the agreement is exempted by the regulation. This procedure was introduced for the first time[77] in the patent licensing regulation, but has not been used as much as expected. This may be because the definitions of the licences to which the regulation applied in article 1(1) were narrow and the black lists of article 3 were long. The latest regulation is of wider application in both respects, so the opposition procedure may be more often useful.

Article 5 prevents the regulation applying to certain arrangements that may be horizontal, and to licences of other kinds of intellectual property rights, unless the latter are ancillary. The new and unusual definition of 'ancillary provision' in article 10(15) of the technology transfer regulation may be important.

Article 6 extends the ambit of the regulations to include sub-licences, assignments, and licensing agreements in which the rights and obligations are assumed by undertakings connected with the parties. This may not be required, in view of the wide concept of 'undertakings' developed by the Court, but does increase certainty.

Article 8 includes various rights less than patents, such as applications, utility models, and supplementary protection certificates for medicinal products which, technically, do not merely prolong the life of the patent. Plant breeders' rights have, at last, been added to the list.

Article 7 provides that the Commission may withdraw the benefit of the block exemptions from agreements that do not merit exemption. Such withdrawal dates only from the decision invoking the article. Similar provisions in all the group exemptions have never been used formally, but the Commission's threat to withdraw a group exemption may cause the parties to amend their agreement.[78]

Article 10 contains the definitions. To qualify, the know-how must be secret, substantial, and recorded, but the definitions of these terms include any technical information for which it is worth paying royalties.

---

[77] There is something rather similar in the group exemptions for transport by rail or inland waterways.

[78] The only case where the Commission has formally withdrawn a block exemption was *Langnese* (93/406/EC), [1993] OJ L183/19, [1994] 4 CMLR 51, [1993] 2 CEC 2123, although it did persuade the parties to abandon an exclusive licence in *Tetra Pak I* (T–51/89), [1990] ECR II–309, [1991] 4 CMLR 334, [1990] 2 CEC 409.

The transitional provisions are in article 11.

Recital 1 no longer states that the communication of know-how is often irreversible and greater contractual certainty is necessary so that each party can expect to appropriate to itself the benefits of its investment. That might have been difficult to reconcile with the Commission's desire to adopt a unifying exempting regulation. It is thought, however, that the recitals omitted from the new regulation remain relevant to provisions that have not been substantially changed. The draftsman was under pressure to reduce the length of the regulation.

4

Underlying Principles and Ancillary Provisions

Title XV of the Maastricht Treaty affirmed the Community's commitment to research and technical development. Article 130(f) of the Treaty on European Union provides:

> the Community shall, throughout the Community, encourage undertakings including small and medium sized undertakings, research centres and universities in their research and technological development activities of high quality: it shall support their efforts to cooperate with one another, aiming, notably, at enabling undertakings to exploit the internal market potential to the full.

Since Maastricht, the Commission has been trying to adopt texts that are both clearer and simpler, even understandable by those who are not specialists. Consequently, the technology transfer regulation has been made simpler than its predecessors. It is also more liberal, and more technology licences can be brought within its terms than under the earlier regulations.

The introductory words of article 1(1) of the technology transfer regulation exempt agreements between only two undertakings whereby patents and or know-how are licensed.

4.1 THE RECITALS

There are fewer recitals in the technology transfer regulation than in the earlier regulations to help with the construction of the provisions in the operative part. But where substantive articles are unchanged the recitals to the former regulation must be relevant and will be considered in this monograph in relation to the provisions they govern.

4.1.1 Harmonize and Simplify the Two Former Regulations

Recital 3 of the technology transfer regulation makes it clear that the exemption extends to patent, know-how, or mixed licences of patents and know-how. It is harmonizing and simplifying the former regulations:

> (3) These two block exemptions ought to be combined into a single regulation covering technology transfer agreements, and the rules governing

patent licensing agreements and agreements for the licensing of know-how ought to be harmonized and simplified as far as possible, in order to encourage the dissemination of technical knowledge in the Community and to promote the manufacture of technically more sophisticated products. In those circumstances Regulation (EEC) No 556/89 should be repealed.

This is done by article 11, which also renews the patent regulation until the new regulation comes into force and provides that agreements that were in force at that time and exempted by the former regulations will continue to be exempt.[1]

4.1.2 Pure Patent, Pure Know-how, and Mixed Patent and Know-how Licences are Exempted (Recital 3)

The new regulation applies to pure patents, pure know-how, or mixed patent and know-how agreements. Recital 4 states that the regulation applies to licensing of national, Community, and European patents, which are called 'pure patent licensing agreements',[2] to the communication of non-patented technical information and to licences combining both elements.

Article 8, described at 5.2 to 5.2.4 below, lists some other intellectual property rights that are deemed to be patents for the purposes of the regulation, and article 6 includes sub-licences, assignments, and agreements when a connected company assumes the rights or obligations.[3] Recital 6 and the introductory words to article 1(1) go further and include agreements including ancillary[4] provisions relating to other intellectual property rights. If these other rights are not ancillary, article 5(1)(4) excludes the application of the regulation.[5]

4.1.3 The Benefits of Territorial Restraints (Recital 12)

The general benefits of exclusive licences supported by limited territorial restraints are described in recital 12 and are important when arguing that an agreement, for instance, of pure software that comes outside the regulation deserves favourable treatment:

> (12) The obligations listed in Article 1 generally contribute to improving the production of goods and to promoting technical progress. They make the holders of patents or know-how more willing to grant licences and licensees

---

[1] See 9.2.1 below.
[2] This corresponds to the first part of recital 4 in the patent reg.
[3] See 5.2–5.2.4 below.
[4] Note the wide definition of 'ancillary provisions' in art. 10(15), considered at 5.1.4 below.
[5] Discussed at 5.1.4 below.

more inclined to undertake the investment required to manufacture, use and put on the market a new product or to use a new process. Such obligations may be permitted under this Regulation in respect of territories where the licensed product is protected by patents as long as these remain in force.

At this point, the Commission is clearly perceiving the need for territorial protection *ex ante*, when licensor and licensee are making commitments to invest.

Recital 1 of the know-how regulation, after mentioning the *vires*, went further in explaining the need for an exemption for know-how agreements: it mentioned the increasing economic importance of know-how, the large number of licensing agreements, that the transfer of know-how is often irreversible in practice, and that, consequently, firms need to be able to rely on their contracts if the technology is to be disseminated.

The absence of intellectual property rights used to be treated as a reason for appraising know-how licences more strictly than patent licences. In the days when the Commission thought that only those provisions in a patent licence that came within the scope of the patent were protected,[6] the only protection the Commission cleared in a know-how licence as being outside the prohibition of article 85(1) was an obligation of confidentiality, which it treated as the specific subject matter of know-how.

By 1989, however, when the know-how regulation was adopted, the Commission accepted that the vulnerability of know-how was a reason for permitting more protection, rather than less, because it was difficult to assure those investing in creating or exploiting it that they would be able to appropriate the results of their investment. This could be done only by contracts which had to be be enforceable. This argument seems to have prevailed even in relation to mixed patent and know-how licences, where vital parts of the technology may be protected by patents. Nevertheless, there had to be substantial additional know-how that remained secret for the know-how regulation to apply.

The absence of these recitals in the technology transfer regulation does not matter where a licence clearly comes within its terms, but if it does not, the question arises whether the Commission's view has changed. In my view, it has not. Recital 3 states that the patent and know-how regulations are to be unified. It does not follow that there are no differences between know-how that is protected by patent and that which is not.

4.1.4 Open Exclusive Licences (Recital 10)

Recital 10 sets out the Commission's understanding of the Court's

---

[6] See **James Venit**, 'In the Wake of *Windsurfing*: Patent Licensing in the Common Market', in **Hawk** (ed.), *Thirteenth Annual Proceedings of the Fordham Corporate Law Institute* (New York: 1986), 517.

judgment in *Nungesser*.[7] Again, it is most useful when trying to persuade a national court or the Commission that an agreement that does not come within the regulation is not forbidden by article 85(1) or that it merits an individual exemption by the Commission. The recital states:

> (10) Exclusive licensing agreements, i.e. agreements in which the licensor undertakes not to exploit the licensed technology in the licensed territory himself or to grant further licences there, may not be in themselves incompatible with Article 85(1) where they are concerned with the introduction and protection of a new technology in the licensed territory, by reason of the scale of the research which has to be undertaken, of the increase in the level of competition, in particular inter-brand competition, and of the competitiveness of the undertakings concerned resulting from the dissemination of innovation within the Community. In so far as agreements of this kind fall, in other circumstances, within the scope of Article 85(1), it is appropriate to include them in Article 1, in order that they may also benefit from the exemption.

This confirms the Commission's acceptance of the Court's judgment in *Nungesser* and the narrow interpretation it applied to the precedent in *Boussois/Interpane*,[8] *Rich Products/Jus-Rol*,[9] and *Delta Chemie*.[10] It extends the Court's ideas only in as much as it applies them to know-how and patents and not only to plant breeders' rights. Since the language of the judgment was based on the need to induce investment and was not expressly limited to plant breeders' rights, the extension is slight, but important.[11]

The reference to the parties' 'competitiveness' was introduced in the know-how regulation. The concept is creeping into Commission documents.[12] This may not always be a reference to the competitive conditions

---

[7] *Erauw-Jacquéry (Louis) Sprl* v. *La Hesbignonne Société Coopérative* (27/87), [1988] ECR 1919, [1988] 4 CMLR 576, [1989] 2 CEC 637, described at 1.4.1, goes further than *Nungesser (L. C.) KG and Kurt Eisele* v. *Commission* (258/78), [1982] ECR 2015, [1983] 1 CMLR 278, CMR 8805, described at 2.2–2.2.5 above, in accepting absolute territorial protection for the propagators of basic seed when required to protect the specific subject matter of plant breeders' rights, the holder may impose customer restrictions to prevent defective handling, and may impose resale prices.

[8] *Boussois/Interpane* [1987] OJ L50/30, [1988] 4 CMLR 124, CMR 10,859. See 3.1.3 above.

[9] [1988] OJ L69/21, [1988] 4 CMLR 527, CMR 10,956, described at 3.1.4 above.

[10] *Delta Chemie/DDD* [1988] OJ L309/34, [1989] 4 CMLR 535, [1989] 1 CEC 2254, described at 3.1.5 above.

[11] See 2.2.4 above. The narrowest construction of the precedent would be limited to open exclusive licences of plant breeders' rights when the technology is new and investments are required by both parties.

It is only for the purpose of the technology transfer reg. that plant breeders' and other intellectual property rights are considered to be patents by virtue of art. 8, described at 5.2.4 below.

[12] The Commission refers in the third para. of the introduction to its *Seventeenth Report on Competition Policy* (Brussels: EC Commission, 1988) to the importance of an internal market

of the market, as much as to the ability of Community firms to compete with firms outside the Common Market. A cynic might assume that it includes any advantage to the parties or disadvantage to their competitors, whether or not it be based on greater efficiency.

In the context of recital 10, however, competitiveness must exclude the grant of state or Community aids to firms that could not compete abroad without technology paid for by an aid. The Commission refers, indirectly, to the need to encourage risky investment if new technology is to compete with old. This sounds like a competitive criterion.

When is technology still 'new' and how new need it be? The Court disagreed with the Commission on the second point in *Nungesser*, discussed at 2.2.3 above. I have been told that a rule of thumb is being applied by some officials, treating know-how as new for about five years, but some practitioners are less sure. In *Rich Products/Jus-Rol* the technology did not count as new because of rival know-how, although this was not stated to be in the public domain. My own view is that in *Nungesser* novelty was only one of the factors that influenced the Court, and was given only as an example, not as a condition. What is important is the need to encourage investment.

4.2 THE REGULATION MAY APPLY WHEN THE TERRITORY OR OBLIGATIONS EXTEND BEYOND THE COMMON MARKET (RECITAL 7)

The Commission has long claimed that agreements, even if made abroad by undertakings based in non-member states, are prohibited by article 85(1) if they have perceptible effects in the Common Market of the kinds thereby prohibited.[13] In *Wood Pulp I*,[14] the Court upheld the Commission's

---

without barriers, and to '[t]he resulting improvement in industrial efficiency and competitiveness and in overall economic performance.' This does not throw much light on the meaning of the concept but does seem to associate it with some concept of efficiency, giving better value to consumers.

In its *Twenty-fifth Report on Competition Policy* (Brussels: EC Commission, 1995), in the first heading of the introduction, the Commission seems to be using the concept in the same way. It published a white paper on *Growth, Competitivity and Employment*, COM(93)700 final, 5 Dec. 1993, highlighting the role of technology transfers in boosting the competitiveness of Community industry and economic growth.

[13] In *BBC Brown Boveri* [1988] OJ L301/68, [1989] 4 CMLR 610, CMR 11,035, the Commission found that an exclusive licence to manufacture in Japan with an implied restriction on selling outside in Member States where there was no patent infringed art. 85(1), although it had held that the licensee would not have been a potential competitor but for the agreement.

In *Raymond/Nagoya* [1972] JO L143/39, [1972] CMLR D45, [1972] CMR 9513, on the other hand, Nagoya was granted an exclusive right to manufacture plastic attachment components perfected by Raymond in Japan and to sell them in various countries in the Far East. Vehicles incorporating such components were permitted to enter the common market, but there was a restriction on exporting the components separately from vehicles. The

power to impose fines on foreigners for implementing within the common market an agreement alleged to have been made outside. The Court applied the objective territorial principle which does not find jurisdiction when the only effects in the common market are indirect. It is not clear whether there is jurisdiction to fine parties for an agreement not to carry on particular types of business within the common market or to prohibit such an agreement. Is it implemented by not operating within the common market? If so, a firm established outside the common market with assets within it might be fined and the agreement rendered void.

The Commission has also decided that restraints on exports outside the common market may infringe article 85(1) if the goods that might otherwise have been exported would have been likely to come back to the common market and be traded over national boundaries and so affect competition and inter-state trade. Less often is it interested in the effects on competition within the common market of a restriction on exporting.[15]

Article 1(1) of the technology transfer regulation exempts specified obligations on the licensor not to manufacture or sell in the licensed territory, or by a licensee not to exploit the licensed technology in

---

Commission cleared even the export ban on the ground that it was unlikely to have perceptible effects on competition within the common market.

Restrictions on conduct outside the Community have been cleared only when, for some reason, it was unlikely that, had the conduct taken place, it would have affected the common market.

In *Junghans* [1977] OJ L30/10, [1977] 1 CMLR D82, CMR 9912, a clause restraining a distributor from selling outside the Community was cleared because the tariffs then payable as the goods left and returned to the common market would render such trade unprofitable. The Commission observed, however, that from July 1977, tariffs between the EC and countries with which the Community has made free trade agreements would cease and warned that such clauses may thereafter infringe art. 85. Compare *David Campari Milano SpA* [1978] OJ L70/69, [1978] 2 CMLR 397, CMR 10,035, described at 3.1.2 above, at para. 60.

[14] *Woodpulp I—A. Ahlström Osakeyhtio and Others* v. *Commission* (89/85), [1988] ECR 5193, [1988] 4 CMLR 901, CMR 14,491. Various producers of wood pulp in countries outside the common market were alleged to have engaged in a concerted practice of selling it at identical prices. The Court decided that the Commission is competent to fine firms that collude outside the common market if they sell into the common market from places outside. The Court's reason is that a contract has two aspects—it may be made and it may be implemented. It stated at para. 16 ff.: '[t]he decisive factor is therefore the place where it is implemented.'

The judges draft their judgments in French and the participle in the French text of the judgment is '*mise en œuvre*'.

'17. The producers in this case implemented their pricing agreement within the Common Market. It is immaterial in that respect whether or not they had recourse to subsidiaries, agents, subagents, or branches within the Community in order to make their contacts with purchasers within the Community.

18. Accordingly the Community's jurisdiction to apply its competition rules to such conduct is covered by the territoriality principle as universally recognised in public international law.'

[15] e.g. the Notice published under art. 19(3) of reg. 17 in *Fluke/Philips* [1990] OJ C188/2, [1990] 4 CMLR 166.

territories within the common market reserved for the licensor or for which
he has granted licences to third parties. Article 10(11) provides that:

> 'the licensed territory' is the territory covering all or at least part of the
> common market where the licensee is entitled to exploit the licensed
> technology;

The territorial protection, limited to territories within the Common
Market, was extended by recital 4 of the know-how regulation, which
largely reproduces the law established in relation to exclusive distribution
agreements in *Hydrotherm*.[16] There, the Court ruled that an exclusive
distribution agreement that extended beyond the common market might
benefit from the exemption granted by regulation 67/67 which also referred
to an exclusive territory within a defined part of the common market.
Advocate General Lenz gave various reasons for this conclusion. He
referred to the positive effects of exclusive distribution agreements and
said that they are often the only means for small and medium-sized firms to
compete, and that small producers are particularly likely to grant
territories that extend beyond the common market. He also pointed out
that there are provisions ensuring that the territorial protection is not
absolute—sub-purchasers are free of restrictions.[17] These points apply
equally to technology licences.

Recital 7 is virtually the same as recitals 4 and 5 in regulation 2349/84
relating to patent licences. It states:

> (7) Where such pure or mixed licensing agreements contain not only
> obligations relating to territories within the common market but also
> obligations relating to non-member countries, the presence of the latter does
> not prevent this Regulation from applying to the obligations relating to
> territories within the common market.
>
> Where licensing agreements for non-member countries or for territories
> which extend beyond the frontiers of the Community have effects within the
> common market which may fall within the scope of Article 85(1), such
> agreements should be covered by this Regulation to the same extent as would
> agreements for territories within the common market. [I have divided a
> single paragraph into two for ease of reference.]

The wording of the first paragraph is limited. It is only the obligations
relating to the common market that are exempted, not those relating to
foreign countries, which may indirectly affect the common market.[18] So a

---

[16] *Hydrotherm Gerätebau GmbH* v. *Compact del Dott. Ing. Mario Andreoli & C Sas* (170/
83), [1984] ECR 2999, [1985] 3 CMLR 224, CMR 14,112.
[17] At 3026 of the ECR.
[18] Since there is a direct effect within the common market, and the parties may well be
established or the agreement made within the common market, there may be no need to rely
on the *Wood Pulp Case*.

restriction on a German or Polish licensee exporting to Hungary would not be covered. Territorial protection might induce the Hungarian licensee to tool up and, eventually, supply a substantial part of the common market requirements, but exclusivity and the associated restraints on making or selling outside the territory are often treated by the Commission as requiring an exemption,[19] and the restriction on selling in Hungary is not covered by the first paragraph.

The second paragraph of the recital presupposes that at least part of the licensed territory is outside the common market, so the obligations relating to Hungary imposed on the Polish licensee may come within the regulation, but not those of the German licensee. If I am right, the situation is anomalous. The Commission seems to think that there is no lacuna. If there is it is a small one, as the restriction on licensees whose territories are confined to the common market not to export to a non-member state will seldom have appreciable effects on competition within the common market that need exemption. The cautious negotiator may extend the territories of licensees within the common market outside to take advantage of the second paragraph if other licensees for territories outside the common market require protection from them. I hope that this artificial course, which may not be convenient for territories where there is no contiguous non-member state, is not necessary.

Unfortunately, the Court's judgment in *Hydrotherm* deals only with the case of a territory including areas outside the common market—there, Western Europe—and does not deal with the protection of a territory entirely outside the common market from competition from a territory entirely within the common market. There is, however, no reason to prevent the Court from extending its judgment to such a case. The reasons given by the Advocate General apply. Willy Alexander construed recital 4 of the patent licensing regulation[20] to mean:

> that obligations as mentioned in the black list of article 3 do not prevent the regulation from applying as long as they relate to non-member countries only.

This view avoids the difficulties. It is possible, however, that the Commission intends to exempt licences to exploit territories outside the common market only to the same extent as those to exploit territories within it. The argument to the contrary is that an exclusive territory outside the common market is likely to have less effect on competition within the Community than such a territory within it. So no exemption is necessary.

---

[19] See 1.1 above.
[20] **Willy Alexander**, 'Block Exemption for Patent Licensing Agreements: EEC Reg. No. 2349/84' (1986) 17 *IIC* 1, 15.

4.2.1 EEA Agreement

Article 53 of the EEA Agreement[21] is very similar to Article 85 (EC) where an agreement 'may affect trade between the Contracting Parties'. Article 56 prescribes the circumstances in which the EC Commission has competence and those agreements over which the EFTA Surveillance Authority (ESA) is competent.[22]

Since the three largest members of the EEA—Austria, Finland, and Sweden—joined the Community at the beginning of 1995, the only non-common market members of the EEA are Norway, Iceland, and Liechtenstein. Consequently, most agreements come within the competence of the EC Commission. Although Switzerland was a member of EFTA, it did not join the EEA. The three countries that did join are normally referred to as 'the EFTA Member States'.

The technology transfer regulation should be incorporated into EEA law by the adoption by the EEA Joint Committee of a decision.[23] That regulation will be expressed to embrace the whole of the EEA area, including the common market and the three EFTA states. When trade between only EC Member States is affected by a licensing agreement, the EC regulation will apply, but when trade with one or more of the EFTA members of the EEA is affected, the EEA regulation will apply whether or not trade between the Member States of the EC be affected.

Given the political determination that the EEA should be kept separate from the EC, this system causes the minimum inconvenience—a single system will embrace either the whole Common Market or, if trade with an EFTA country be affected, EEA law. Marginal problems are not serious, even if it is not clear whether trade with the non-EC Member States is affected sufficiently for the EEA law to be relevant. Usually it will not matter whether the EC or EEA regulation is applicable since the substantive law will correspond. There should be no problems deciding

---

[21] I am indebted to Helge Stemshaug of the University of Oslo for help with an earlier draft of this sect. Many thanks! Naturally, I remain responsible for any errors remaining.

[22] The ESA is competent where trade only between the EFTA members is affected or where the turnover of the parties in the EFTA states represents 33% or more of their turnover in the territory. Where the conduct has little effect on trade between the EC Member States or on competition within the Community, the ESA is competent. Complaints may be made to either surveillance authority.

Art. 58 and Protocol 23 require the two surveillance authorities to consult each other. Art. 10(2) of Protocol 23 provides that if notifications or complaints are sent to the wrong surveillance authority, they shall be sent without delay to the other. Consequently, if the parties work out the turnover for the two territories or other criteria wrongly, it should not matter.

[23] After the EEA reg. is adopted by the Joint Committee, it will have to be incorporated into the national legislation of the 3 EFTA Member States.

whether the ESA or EC Commission is competent to oppose the application of the group exemption or withdraw the benefit of the exemption under article 7 because of the duty of each surveillance authority to pass on documents to the other.[24] Moreover, the application of the regulation is likely to be decided mainly by national courts and it will not then matter which regulation applies.

The EFTA Court has been careful to follow the case law of the Community Court.[25] So for substantive purposes it may not matter to which court an appeal can be made. Even if there is difficulty in deciding which surveillance authority is competent in marginal cases, this should not affect the parties.

The doctrine of exhaustion of intellectual property rights under the rules for the free movement of goods, which was explained at 2.1.2 to 2.1.3, applies throughout the EEA,[26] but not between Switzerland and the Common Market or under the Europe agreements under which some East European countries may later join the Common Market.

Since the EEA regulation has not yet been adopted, I will refer in this book to the EC regulation and to the Common Market. This should be understood to embrace the whole European Economic Area and its regulation where relevant.

4.2.2 Free Trade Agreements

Any restrictions on sales from Switzerland to Germany may be subject also to the provisions protecting competition in the territories of the free trade agreement (FTA) made between Switzerland and the Communities. The provision reads very like article 85, but the sanctions for infringing it are weak: diplomatic complaints may be made to a Joint Committee set up

[24] Protocol 23, as explained in n. 22 above.
[25] The cases did not relate to the competition rules.
   Art. 6 of the EEA Agreement provides that provisions that are substantially identical to those in the EC and other treaties shall be interpreted in accordance with Community case law as it stood in Aug. 1992.
   Art. 3(2) of the EFTA Agreement requires the ESA and EFTA Court to pay due account to the principles and rulings laid down by the Community Court thereafter. *Carte blanche* has not been given to the Community Court.
[26] See Protocol 28 to the EEA Agreement on Intellectual Property, art. 2:
   '*Exhaustion of rights*
   1. To the extent that exhaustion is dealt with in Community measures or jurisprudence, the Contracting Parties shall provide for such exhaustion of intellectual property rights as laid down in Community law. Without prejudice to future developments of case law, this provision shall be interpreted in accordance with the meaning established in the relevant rulings of the Court of Justice of the European Communities given prior to the signature of this agreements.
   2. As regards patent rights, this provision shall take effect at the latest one year after the entry into fore of this agreement.'

under the Agreement. The Community Court decided in *Wood Pulp I*[27] that the provisions in the free trade agreement with Finland before its accession to the Communities that protect competition:

> presuppose that the Contracting Parties have rules which enable them to take action against agreements which they regard as being incompatible with that Agreement. As far as the Community is concerned, those rules can only be the provisions of Articles 85 and 86 of the Treaty. The application of those articles is therefore not precluded by the Free Trade Agreement.

The provisions of the free trade agreements may have direct effect,[28] but the direct effect does not incorporate the provision into the law of the Member States since free trade agreements do not create a common market, but only a free trade zone.[29]

There is no question of a group exemption from the competition rules of the free trade agreements. So recital 7 applies to such contracts in the same way as to imports from other states that are not members of the EEA.

4.2.3 The Europe Agreements between the EC and Some States of Eastern Europe

The Europe Agreements also contain provisions modelled on articles 85 and 86, but they do not have even a remote direct effect.

As the East European economies are opening up, licensors to the Eastern block are increasingly concerned about the validity of export bans. Prices may have to be far lower in Eastern Europe than in the Common Market and it may not be worth selling or licensing into Eastern Europe if the products cannot be kept out of the Common Market.

The Community doctrine of exhaustion[30] does not apply on the first sale into the Common Market, so intellectual property rights may be used to keep out goods sold in Eastern Europe with the licensor's consent. Moreover, it seems from article 2(1)(14) of the technology transfer regulation, described at 7.3.14 below, that a licence to produce and sell in one country does not exhaust patent rights in a Member State by Community law. This may also be relevant to other rights intended to

---

[27] N. 14 above, at para. 31.

[28] *Hauptzollamt Mainz* v. *C. A. Kupferberg & Cie KG* (104/81), [1982] ECR 3641, [1983] 1 CMLR 1, CMR 8877.

[29] *Polydor Ltd and RSO Records Inc.* v. *Harlequin Record Shops Ltd* (270/80), [1982] ECR 329, [1982] 1 CMLR 677, CMR 8806. The judgment related to rules for the free movement of goods, but it is thought that the competition rules are even less likely to be incorporated into the rules of Community law. For earlier views, see **Neville March Hunnings**, 'Enforceability of Free Trade Agreements' (1977) 2 *ELR* 163, with reply, by **Michel Waelbroeck** 'Enforceability of the EEC–EFTA Free Trade Agreements: A Reply' (1978) 3 *ELR* 27, and rejoinder, *ibid*. 278.   [30] Described at 2.1.2–2.1.4 above.

encourage investment, such as copyright or design rights. The law on trade marks may be different.

International exhaustion under national law is also being discouraged by the directives harmonizing intellectual property.[31] So holders may be able to rely on their intellectual property rights to keep goods sold or made under licence in East Europe.

Limited contractual restraints may also be possible. The second paragraph of recital 7 of the technology transfer regulation applies the regulation to restrictions on active and passive sales to Common Market territories from outside them and the EEA regulation will cover such sales to the EFTA countries.[32]

[31]The Commission seems to accept that restraints on imports from East European countries may be necessary where products are sold at little more than marginal cost there. For political reasons, however, it hesitates to express this publicly. Fear of such imports may be the reason that the dirs. on intellectual property law are being construed by the Commission to restrain Member States from adopting a doctrine of international exhaustion. See, e.g., the first Council dir. 89/104 of 21 Dec. 1988 to approximate the laws of the Member States relating to trade marks, [1989] OJ L40/1, art. 7, described at 3.3 above.

[32] Neither applies to sales into another East European country where an intermediate trading post may be set up. Some contracts require an East European licensee not to sell 'directly or indirectly' into the EEA to the extent that export bans are permitted by the reg., but the word 'indirectly' is dangerous, as it may cover parallel importers generally and not only the establishment of a trading post.

5

Exclusions and Inclusions

Various kinds of agreement are excluded from the application of the exemption by article 5(1).

5.1.1 Horizontal and Reciprocal Agreements (Recital 6 and Article 5(1)(1)–(3))

As in the patent and know-how licensing regulations, article 5 of the technology transfer regulation excludes various horizontal and reciprocal agreements. The recitals to the patent regulation state that the Commission has had insufficient experience of these kinds of transaction. This is not stated in the later regulations, but there have been some cases, not all of them before the patent licensing regulation was adopted in 1984, which are considered in 5.1.1.1 to 5.1.1.1.3 below. The final words of recital 8 indicate that in appropriate cases, an individual exemption might be granted:[1] '[s]uch agreements pose different problems which cannot at present be dealt with in a single regulation (Article 5).'

Blanket *per se* legality would not be appropriate for patent pools or joint ventures, but exemption or clearance may be sensible in particular cases in the absence of market power or where efficiencies outweigh any anti-competitive effect. The later parts of recital 8 govern article 5(1) which, in the first three paragraphs, lists various kinds of horizontal and reciprocal agreements.

The exemption in article 1(1) is not limited to agreements between those who could not have competed but for the technology licence. It may, therefore, exempt some horizontal licences between competitors. Article 5(1) excludes some of these from the scope of the regulation, but is not focused solely on horizontal agreements. Patent pools are excluded, even if the parties do not compete and each contributes complementary technology

---

[1] See also the notice under art. 19(3) in *Fluke/Philips* [1989] OJ C188/2, [1990] 4 CMLR 166, and the Commission's *Nineteenth Report on Competition Policy* (Brussels: EC Commission, 1990), point 47, where the Commission approved of a reciprocal exclusive distribution agreement by comfort letter, although the parties competed in relation to a small part of the products affected by reciprocal exclusive territories, a condition blacklisted by art. 3 of reg. 1983/83. There were efficiencies expected from the agreement and it was not intended to keep Fluke out of the Europe. It had tried to enter it with little success over several years.

to which the other or others have no access. Article 5(1)(3) excludes cross-licences where the parties are competitors in relation to the products covered by the agreements, but not unilateral exclusive licences granted to a competitor. The wording is virtually identical in all three regulations.

The relationship between the first three sub-paragraphs of article 5(1), technology pools, reciprocal exclusive rights, and licences affecting joint ventures is not clear. They embrace overlapping concepts, to which different rules apply. The US agencies treat a patent pool and reciprocal rights in a similar fashion in the same section of their guidelines without distinguishing between them.[2] Agreements between the members of a technology pool are excluded from the group exemption, whether or not the parties competed with each other before the licence was granted; while licences to a joint venture or reciprocal exclusive rights are excluded only where the parties competed. A 'patent or technology pool' is not a term of art; it cannot always be distinguished from a joint venture, and seldom from the reciprocal licences excluded by article 5(1)(3).

In its *Eleventh Report on Competition Policy*, points 92 to 94, before a distinction between patent pools and reciprocal exclusive rights had been drawn, the Commission used the term 'patent pool' to include the bringing of patents of two firms together so that they may jointly be made available for use by both parties for their joint benefit. It described an informal decision in *Concast/Mannesmann*,[3] where two firms made all their patents relating to continuous castings available to each other, but exploited them independently.

When they sued jointly for infringement, the defendant complained to the Commission, which stated that: (1) the patent pool eliminated potential competition between the parties as regards technical innovation, and (2) the concentrated technology encouraged customers to deal with those two firms, which supplied some 60 per cent of an important market, rather than others. It closed its file only after the firms terminated their co-operation in 1981.

5.1.1.1 Patent pools and reciprocal agreements (article 5(1)(1) and (3))

Article 5 provides that:

1. This regulation shall not apply to:[4]

    (1) agreements between members of a patent or know-how pool which relate to the pooled technologies;

---

[2] Sect. 5.5, reproduced at 5.1.1.1.1 below.

[3] *Eleventh Report on Competition Policy* (Brussels: EC Commission, 1983), point 93.

[4] Note that art. 5(2) provides that the reg. does apply to these agreements if there is no territorial protection. See 5.1.1.1.3 below.

. . .

(3) agreements under which one party grants the other a patent and/or know-how licence and in exchange the other party, albeit in separate agreements or through connected undertakings, grants the first party a patent, trademark or know-how licence or exclusive sales rights, where the parties are competitors in relation to the products covered by those agreements;

One would not expect patent pools and reciprocal licences to be exempted *en bloc*. They may be efficient in enabling members to develop technology, separately or in collaboration, without worrying whether they are infringing each other's rights, especially where they contribute complementary resources. Some patent pools may be exempt under regulation 418/85, which applies to collaboration for research and development between firms that do not compete or whose aggregate market shares in the common market do not exceed 20 per cent although it seldom applies when the parties need the results to compete in the same markets, for reasons I have tried to analyse elsewhere.[5] Where the parties make complementary contributions, or where collaboration does not extend beyond research and development as widely defined in that regulation, the agreement probably does not infringe article 85(1)[6] and, if it does, the Commission is likely to grant an individual exemption or send a comfort letter closing the file.

On the other hand, as the Commission alleged in *Concast Mannesmann*, 5.1.1 above, technology pools may exclude third parties and discourage competition in research and development between the parties. They may be naked cartels between competitors, ancillary to no agreement to carry on research and development more effectively, and there may be no need to avoid blocking patents, in which case, they should be prohibited. Moreover, when two of only a few independently viable technologies are brought together by the acquisition of exclusive rights to the second, patent

---

[5] I spoke on the subject to LES Scandinavia in Sept. 1995 and my speech is to be published in *Les Nouvelles*. See also, **Valentine Korah**, 'Research and Development, Joint Ventures and the European Economic Community Competition Rules' (1988) 3 *Int. J Technology Management* 7.

[6] Recital 2 of reg. 418/85 states that:

'As stated in the Commission's 1968 Notice concerning agreements, decisions and concerted practices in the field of co-operation between enterprises, agreements on the joint execution of research work or the joint development of the results of the research, up to but not including the stage of industrial application, generally do not fall within the scope of Article 85(1) of the Treaty. In certain circumstances, however, such as where the parties agree not to carry out other research and development in the same field, thereby foregoing the opportunity of gaining competitive advantages over the other parties, such agreements may fall within Article 85(1) and should therefore not be excluded from this Regulation.'

pools may exclude third parties. In its Christmas message of 1962,[7] the Commission wisely refused to make generalizations about patent pools.

5.1.1.1.1 US Government Policy Towards Patent Pools

The *Antitrust Guidelines for the Licensing of Intellectual Property*[8] announced by the US Department of Justice and the Federal Trade Commission in 1995 observe at 5.5 that cross-licensing and pooling arrangements:

> may provide pro-competitive benefits by integrating complementary techno-logies, reducing transaction costs, clearing blocking positions, and avoiding costly infringement litigation. By promoting the disseminating of technology, cross-licensing and pooling arrangements are often pro-competitive.
>
> Cross-licensing and pooling arrangements can have anti-competitive effects in certain circumstances. For example, collective price or output restraints in pooling arrangements, such as the joint marketing of pooled intellectual property rights with collective price setting or coordinated output restrictions, may be deemed unlawful if they do not contribute to an efficiency-enhancing integration of economic activity among the participants. Compare *NCAA* 468 US at 114 (output restriction on college football broadcasting held unlawful because it was not reasonably related to any purported justification) with *Broadcast Music*, 441 US at 23 (blanket license for music copyrights found not per se illegal because the cooperative price was necessary to the creation of a new product). When cross-licensing or pooling arrangements are mechanisms to accomplish naked price fixing or market division, they are subject to challenge under the per se rule. See, *United States* v. *New Wrinkle, Inc.*, 342 US 371 (1952) (price fixing).
>
> Settlements involving the cross-licensing of intellectual property rights can be an efficient means to avoid litigation and, in general, courts favor such settlements. When such cross-licensing involves horizontal competitors, however, the Agencies will consider whether the effect of the settlement is to diminish competition among entities that would have been actual or likely potential competitors in a relevant market in the absence of the cross-licence. In the absence of offsetting efficiencies, such settlements may be challenged as unlawful restraints of trade. Cf. *United States* v. *Singer Manufacturing Co.*, 347 US 174 (1963) (cross license agreement was part of a broader combination to exclude competitors).
>
> Pooling arrangements generally need not be open to all who would like to join. However, exclusion from cross-licensing and pooling arrangements among parties that collectively possess market power may, under some

---

[7] JO 2921/62, retracted when the patent licensing reg. was adopted, much of it quoted at 3.2.1 above.

[8] Issued by the agencies on 6 Apr. 1995, and reproduced in CCH 4 *Trade Regulation Reporter*, ¶13,132, and in (1995) 7 *EIPR Supp.* 3. I described their general approach at 1.4.3 above.

circumstances, harm competition. Cf. *Northwest Wholesale Stationers, Inc.* v. *Pacific Stationery and Printing Co.*, 472 US 284 (1985) (exclusion of a competitor from a purchasing cooperative not per se unlawful absent a showing of market power). In general, exclusion from a pooling or cross licensing arrangement among competing technologies is unlikely to have anti-competitive effects unless (1) excluded forms cannot effectively compete in the relevant market for the good incorporating the licensed technologies and (2) the pool participants collectively possess market power in the relevant market. If these circumstances exist, the Agencies will evaluate whether the arrangement's limitations on participation are reasonably related to the efficient development and exploitation of the pooled technologies and will assess the net effect of those limitations in the relevant market. See section 4.2.

Another possible anti-competitive effect of pooling arrangements may occur if the arrangement deters or discourages participants from engaging in research and development, thus retarding innovation. For example, a pooling arrangement that requires members to grant licenses to each other for current and future technology at minimal cost may reduce the incentives of its members to engage in r & d because members of the pool have to share their successful r & d and each of the members can free ride on the accomplishments of other pool members.[9] However, such an arrangement can have pro-competitive benefits, for example, by exploiting economies of scale and integrating complementary capabilities of the pool members, (including the clearing of blocking positions), and is likely to cause competitive problems only when the arrangement includes a large fraction of the potential r & d in an innovation market.

5.1.1.1.2 EC Case Law

A few examples may illustrate the concern in the EC. The case law dates from the 1970s and 1980s; there are no recent public examples.

In *Bronbemaling* v. *Heidemaatschappij*,[10] opposition proceedings in the Dutch patent office relating to a patent for a horizontal well-point drainage system were settled by granting licences to the three firms opposing the grant of the patent on terms that, without the consent of a majority, licences would be given to no-one but them and to a subsidiary of the applicant for the patent. The parties were already competitors when the licences were granted. Those opposing the grant of a patent argued that the invention was not novel and that they had all been using it. The invention was important for construction in the Netherlands, much of which lies below sea level, since public authorities and large firms frequently specified use of a well-point draining system.

Bronbemaling and another firm were refused licences and Bronbemaling

---

[9] Citation of case omitted.
[10] [1975] OJ L249/27, [1975] 2 CMLR D67, CMR 9776.

complained to the Commission. The Commission adopted a decision under article 15(6) of regulation 17 withdrawing the immunity from fines from the compromise of the dispute in the patent office that had been notified to the Commission.

There are conflicting considerations as to what should be done concerning settlements of disputes about the validity of a patent, whether at the administrative level or in proceedings before a court. On the one hand, the parties both want to use the patented invention, so are likely to be competitors, and an arrangement to license only the objectors may be a cosy cartel to exclude others. On the other hand, it is desirable to save the costs of litigation or administrative proceedings. The solution should probably be based on the doctrine of proportionality and depend on the appropriateness of the restriction accepted to make dispute-resolution commercially possible.

The Commission seems to have treated the agreement as creating a bottleneck monopoly as some important construction contracts stipulated for the patented system. The only clause to which the Commission objected was that providing limited exclusivity.

It is not clear whether such a case would also have concerned the American enforcement body. It excluded third parties from a market in which the parties to the agreement seem to have enjoyed some market power, but it is not clear how much technology was contributed by the firms opposing the grant of the patent. There was a provision for exchanging know-how. If the licence was purely unilateral it would hardly have amounted to a patent pool, and the exclusion of Bronbemaling would have had no more effect than the exercise of Heidemaatschappij's patent rights, the possibility of which may have induced the original innovations.[11] If the patent was indeed invalid, the compromise may have prevented the competing firms most interested in challenge from opening up the technology for others.

In *Bayer and Hennecke* v. *Süllhöfer*,[12] after a dispute about validity, cross licences were granted, one of them containing a no-challenge clause. The Commission argued[13] that such a clause relating to the disputed technology was automatically legal as being ancillary to the settlement

---

[11] The classic article on the essential facility doctrine is by **Philip Areeda**, 'Essential Facilities: An Epithet in Need of Limiting Principles' (1990) 58 *Antitrust LJ* 841. He argues that the doctrine should be narrowly applied for this reason, and explains the original *Terminal Railroad* case *US* v. *Terminal RR Ass'n.*, 224 US 383 (1912) on the basis that a group of railroads had bought an existing facility, and would not be greatly inconvenienced by making it available to outsiders who wanted to compete with them. He was much more concerned about the grant of a compulsory licence when a single firm had itself invested in creating an essential facility.

[12] (65/86), [1988] ECR 5249, [1990] 4 CMLR 182, [1990] 1 CEC 220.

[13] *Ibid.*, para. 13.

provided the dispute was not a sham, but the Court rejected this view and stated that an agreement to resolve a dispute over intellectual property rights should be appraised in its economic context like any other agreement. It stated that a no-challenge clause may, in the light of the legal and economic context, restrict competition contrary to article 85(1). It suggested that it would not restrict competition when the licence was royalty free or when the invention was technically outdated and not being used by the licensee. Whether it did restrict competition would depend on the market position of the parties.

The Court did not analyse the benefits and anti-competitive effects of an agreement to avoid litigation with any precision, but left the assessment of the legal and economic context to the national court which, in proceedings under article 177 of the EC Treaty, is required to apply the law to the facts.

In *IGR Stereo Television*,[14] there was no dispute. Two German television companies set up a body which obtained patents on inventions needed for the manufacture of a converter or television set which can receive stereo transmissions. In 1980, this body assigned these rights to IGR, whose members included all the firms making colour televisions sets in Germany. IGR granted licences to its members, but resolved to license non-members only from 1983 and only for a limited number of sets. IGR invoked these patents to restrain Salora, a Finnish manufacturer already operating in the Federal Republic, from fulfilling orders for such sets.

Salora complained and the Commission considered ordering interim measures. At that point, IGR agreed to license outsiders immediately. The Commission seems to have thought that the members of IGR had together bought up patents and formed a bottleneck monopoly for the benefit of its members. Unlike joint ventures for r & d, the group exemption of which applies only if the results are pooled, there was no pro-competitive activity to which the pool was ancillary, and a licence required by the competition authorities is an appropriate way to stop such a bottleneck excluding others.

Contrast *Transocean Marine Paint*,[15] an association whose trade-mark licences have been exempted four times on the ground that these were ancillary to the exploitation of a formula for marine paint. Its marketing required a network extending to many countries as ships need paint of the same chemical formula wherever they are maintained.

---

[14] *Eleventh Report on Competition Policy*, n. 3 above, at point 63. In its *Fourteenth Report* (Brussels: EC Commission, 1985), at point 76, the Commission reports that it continued proceedings to the stage of issuing a statement of objections alleging that the patent pool infringed art. 85(1) and had appreciable effects because of the high royalty rates. When these were substantially reduced, the Commission closed its file.

[15] *Transocean Marine Paint Association* [1988] OJ L351/40, [1989] 4 CMLR 621, [1989] 1 CEC 2003, described at 3.1.6 above.

5.1.1.1.3 Under the Technology Transfer Regulation

The regulation distinguishes technology pools from reciprocal rights, but does not indicate how these concepts are to be distinguished.

Sub-paragraph (3) of article 5(1) relating to reciprocal rights is more limited than sub-paragraph (1) for patent pools, in that it applies only where the parties are competitors in relation to the products covered by the agreement. The Commission has drawn a distinction that was not made in its case law before the patent regulation. It is therefore necessary to be able to distinguish technology pools from reciprocal rights.

Which sub-paragraph applies cannot depend on whether more than two firms benefit from the others' technology: the exemption applies only when there are no more than two undertakings party to the agreement. When the making of each agreement is dependent on the others, the network of agreements will form part of a single agreement.[16]

Paragraph (3) speaks in the singular of one party granting to the other. It is true that a licence to a trade association by each of its members comes literally within these words, but there would be at least an expectation that each member would benefit from the technology of all the others and that its technology would be made available to all. This would be part of the same agreement, and so bring the arrangement out of article 1(1) of the regulation, preventing it from applying quite apart from article 5(1). There is no basis for reconciling the provisions when the parties were not competitors in relation to the contract products. It may be helpful to refer to the provision as the grant of reciprocal rights, rather than as an intellectual property pool.

It seems, then, that article 5(1)(1) and (3) excludes technology pools and the reciprocal grant of rights, exclusive or otherwise, from the regulation, even if the parties exploit the technology independently of each other in different product markets and even if the pool creates efficiencies by avoiding problems of blocking patents. These are not grounds of *per se* legality under the regulation, but the efficiencies may be argued before the Commission whether it be acting on a complaint or a request for clearance or exemption. If the arrangement is ancillary to collaboration in research and development, it may fall within the group exemption for joint ventures granted by regulation 418/85.[17]

---

[16] *BP Kemi/DDSF—Atka A/S* v. *BP Kemi A/S and A/S De Danske Spritfabrikker* (79/934/EEC), [1979] OJ L286/32, [1979] 3 CMLR 684, CMR 10,165.

[17] Few agreements between parties who need the results for similar purposes can be brought within this reg., because of the difficulty of ensuring that each can appropriate the agreed share of benefits of its contribution.

The exclusions of article 5(1)(1) and (3) are qualified by article 5(2):

> 2. This Regulation shall nevertheless apply
> (2) . . . to agreements to which paragraph 1(1) applies and to reciprocal
> licences within the meaning of paragraph 1(3), provided the parties are
> not subject to any territorial restriction within the common market with
> regard to the manufacture, use or putting on the market of the licensed
> products or to the use of the licensed or pooled technologies.

The question arises of the scope of the concept of 'any territorial restriction within the Common Market with regard to the manufacture, use or putting on the market of the licensed products or on the use of the licensed or pooled technologies' in Article 5(2)(2). It uses the term 'parties' in the plural. What is the position where only the licensee accepts a restriction? The Commission tends not to distinguish the singular from the plural very carefully. A sole territory does not restrict the licensor in the ways mentioned in article 5(2)(2); an exclusive territory does, however, limit territorially the places where the licensor may put the product on the market,[18] as do the other bans permitted by article 1(1)(2)–(6). The provisions listed in article 2 seldom need an exemption.

Although a sole and exclusive territory does not prevent a licence from being open, the parties might not have made significant investments in developing the technology, preparing to produce and developing a market without such protection, so it may not be caught by article 85(1) and need no exemption even if the parties are competitors, provided that their market share is small.

At first sight, it may seem surprising that the Commission seems to permit the group exemption to apply automatically to technology pools and reciprocal licences where the parties are competitors and the pool will permit prices or royalties to be raised, provided there are no territorial restraints. It must be remembered, however, that for the regulation to apply the agreement must contain one of the provisions listed in article 1(1) and points (2)–(6) of these confer territorial protection. Article 5(2)(2) exempts only a sole, but non-exclusive licence, or a non-exclusive, licence including an obligation to use the licensor's mark or get up, or an obligation to limit production to the licensee's requirements to manufacture its own products.

Moreover, agreements allocating markets by selecting customers or quotas and determining prices are blacklisted, so there may not be an important gap in the exception to the exemption.

'Reciprocal sales rights' in article 5(1)(3) presumably refers to A

---

[18] Sebastiano Guttuso raised the question, without settling it, at the conference organized in Brussels by Legal Studies and Services in Apr. 1989. I had previously assumed that art. 1(1)(1)–(6) was excluded.

granting B rights under a distribution or franchising agreement and B granting A a technology licence. It seems that neither grant need be exclusive for the regulation to be excluded, although an individual exemption or comfort letter would be likely where they are not exclusive. Reciprocal exclusive distribution agreements between competitors is limited by article 3(a) of regulations 1983/83 and 1984/83. Article 5(1)(3) of the technology transfer regulation ensures that these provisions are not circumvented when each grants the other exclusivity under a different regulation but it goes further and covers non-exclusive rights.

The provision for agreements between competitors in the franchising regulation is quite different. Article 5(a) prevents the regulation from applying when actual competitors enter into a franchising agreement in respect of such goods or services, whether or not the agreement is reciprocal. This may be because franchise agreements are treated as normally being vertical, whereas this is not always the case for technology licences.

These exclusions do not apply to licences granted by a patent pool to outsiders or vice versa. Indeed, as we saw in relation to *IGR* the Commission may insist on such a licence being granted on terms it considers reasonable.

Recital 8 excludes article 5(1)(3) and permits the limited territorial protection exempted in article 1 when the reciprocal licences relate to improvements or new applications discovered by the licensee.[19] This is important, as most licences include a feed- or grant-back clause. Few licensors would grant a licence if they had to pass on any improvements they discovered, but could not obtain improvements made by their licensees. They cannot afford to be left behind in the technological race.

5.1.1.2 *Licences between Competitors in a Joint Venture (Article 5(1)(2))*

The exclusion from the exemption of licences between joint venturers or one of them and the joint venture, somewhat like that for reciprocal licences, applies only where the venturers are competitors and the licensing agreement relates to the activities of the joint venture. The exemption does not apply where:

> (2) Licensing agreements between competing undertakings[20] which hold
> interests in a joint venture, or between one of them and the joint

---

[19] Contrast art. 2(1)(4) which whitelists non-exclusive grant-back provisions when the licensor agrees to pass on his improvements with art. 3(6) which blacklists an agreement by the licensee to assign his improvements in whole or in part. These are considered at 7.3.4–7.3.4.4 below.

[20] Contrast art. 5(1)(3) where reciprocal agreements are excluded where the parties are competitors.

venture, if the licensing agreements relate to the activities of the joint
venture;

'Joint venture' is not a term of art in EC law.[21] It does not presuppose a
jointly owned subsidiary, but may consist merely of a joint committee to
which the parents allocate various resources.[22] Where the parties were not
even potential competitors before the agreement, it is thought that the
regulation may apply, even if they subsequently become competitors as a
result of the transaction.[23]

The question arises whether the term 'competitors' includes potential
competitors. 'Competing manufacturers' and 'manufacturers of competing
products' are defined in article 10(17) of the technology transfer regulation
as:

> manufacturers who sell products which, in view of their characteristics, price
> and intended use, are considered by users to be interchangeable or
> substitutable for the licensed products.

Article 5(1)(2) is virtually identical to the equivalent provision in the
earlier regulations. There is, however, no definition of 'competing
undertakings'. It could be argued that this term is wider: the parties need
not already be manufacturing. This, however, would lead to considerable
uncertainty, which would be undesirable in a block exemption. Accordingly, I hope it prevents the exemption applying only when the parties are
actual competitors.

In its *Thirteenth Report on Competition Policy*,[24] when the Commission
made its important statement about joint ventures, and again in the
Commission's notice concerning the assessment of co-operative joint
ventures pursuant to article 85 EEC,[25] it stated that in assessing joint
ventures it would consider all factors relevant not only to actual, but also

---

[21] Save under art. 3(2) of the merger reg., which is irrelevant in this context.

[22] e.g., *GEC/Weir Pump* [1977] OJ L327/26, [1978] 1 CMLR D42, CMR 10,000.

[23] See, more generally, 1.3.1, n. 44 above, where I have given examples from the reg. of
situations where the Commission looks *ex ante* to the position before the agreement is
concluded in assessing the effects of contractual provisions on competition, and 1.4.2.
In *Michell Cotts/Sofiltra* [1987] OJ L41/31, [1988] 4 CMLR 111, CMR 10,852, the
Commission accepted that Mitchell Cotts could not have produced the sophisticated filter
papers by itself and *Optical Fibres* [1986] OJ L236/30, CMR 10,813, mentioned at 1.3.1, n. 44,
where, at para. 46, the Commission stated that since the cable-making party could not have
produced optical fibres without a licence, each joint venture was not, in itself, anti-
competitive. It was only the horizontal effects of a network of joint ventures that required
exemption. But contrast *BBC Brown Boveri* [1988] OJ L301/68, [1989] 4 CMLR 610, CMR
11035, discussed at 1.3.1, n. 44 above, and many other decision on joint ventures such as
KSB/Lowara/Goulds/ITT [1991] OJ L19/25, [1992] 5 CMLR 55, [1991] 1 CEC 2009 even in
recent years.

[24] (Brussels: EC Commission, 1984), for 1983, at the end of 50.

[25] [1993] OJ C43/3 [1996] 5 CMLR 357 at para. 19.

potential, competition. So it is possible that potential competitors are included. The Commission's statements of policy[26] have defined potential competition in a realistic fashion, and it will consider whether more than one party was likely to enter the market without the other.

In the context of the exclusion of patent pools whether or not the members compete, 'joint venture' may be limited to a joint venture to achieve something positive, such as collaborative research and development, and exclude naked restraints where there is no integration between the firms even in relation to the licensed technology, or there would be little to which the first sub-paragraph of article 5(1) could apply.

A joint venture agreement likely to promote competition might well be held by the Court not to infringe article 85(1) if the exclusivity and associated territorial protection were necessary to make it viable.[27] Until now, however, the Commission has tended to exempt exclusivity and not clear it.

In *Mitchell Cotts/Sofiltra*, the Commission cleared a covenant by the parties not to compete in manufacture or sales with their joint venture to which an exclusive pure know-how licence was granted by Sofiltra when Mitchell Cotts could not compete in manufacture, but did compete in selling the product derived from the licence. Would such an agreement come within the transfer of technology regulation if the non-competition clause was changed[28] to an obligation to use the licensee's best endeavours to exploit the licensor's technology, given that at the time the joint venture was created, Mitchell Cotts was already distributing the product which it bought from Sofiltra? Possibly not.

Article 5(1)(2), like the provisions for technology pooling and reciprocal exclusive rights, is qualified by article 5(2):

> 2. This Regulation shall nevertheless apply:
> (1) to agreements to which paragraph 1(2) applies, under which a parent undertaking grants the joint venture a patent or know-how licence, provided that the licensed products and the other goods and services of the participating undertakings which are considered by users to be interchangeable or substitutable in view of their characteristics, price and intended use represent:
>
> — in case of a licence limited to production, not more than 20%, and
> — in case of a licence covering production and distribution, not more than 10%;

---

[26] Unlike many of its decisions granting individual exemption. Some are cited in n. 23 above.

[27] See 1.4.1 above. The Court has not yet given judgment on the merits of co-operative joint ventures.

[28] See 7.3.9–7.3.9.4 for the position under the technology transfer reg.

of the market for the licensed products and all interchangeable or substitutable goods and services;

This provision repeats the provisions of regulation 151/93[29] which amended the group exemptions for joint ventures, patent, and know-how licences to exempt provisions for joint sales where the parties competed, but did not enjoy an aggregate market shares of over 10 per cent, and for joint production without joint sales where their market share did not exceed 20 per cent.

5.1.1.3 Inter-relationship between article 5(1)(1)–(3)

It is not always possible to distinguish between patent pools, the grant of reciprocal exclusive rights and joint ventures. As suggested at 5.1.1.1 above, it would be desirable for article 5(1)(1) to be interpreted narrowly, so as to leave some scope for article 5(1)(2) and (3) which apply only when the parties are competitors in relation to the products covered by the exclusive rights, but this is far from clear. Commission officials often repeat the Court's view[30] that exemptions are acts of policy to be treated as derogations from Article 85 and interpreted narrowly. An exception from an exemption should, therefore, be treated broadly. Nevertheless, the Court has interpreted regulation 67/67 quite broadly.[31]

Article 5(1)(3) relating to reciprocal rights is more liberal than article 5(1)(2) relating to joint ventures also in that it excludes the exemption only if the parties are competitors in relation to the products covered by the

---

[29] [1993] OJ L21/8.

[30] See *Stergios Delimitis* v. *Henniger Bräu AG* (C–234/89), [1991] ECR I–935, [1992] 5 CMLR 210, [1992] 2 CEC 530, opinion of Van Gerven AG, para. 5:

'[I]t is not for the national court, or for the Court of Justice in the context of a reference for a preliminary ruling, to alter or add to the contents of a generic exemption issued by the Commission. The issue of such an exemption is an act of policy which falls within the exclusive competence of the Commission. Consequently, when an agreement is not covered by the terms of a block exemption regulation, that block exemption, in itself a derogation from the prohibition under Article 85(1), and therefore to be strictly interpreted, may on no account be extended.'

This statement was not repeated in the judgment, but the Court interpreted the regulation literally and did not include in it an agreement that did not come within the express words, despite encouragement to do so from the Commission's counsel.

[31] e.g., in *Hydrotherm Gerätebau GmbH* v. *Compact de Dott. Ing. Mario Andreoli & C Sas* (170/83), [1984] ECR 2999, [1985] 3 CMLR 224, CMR 14,112, the Court ruled that reg. 67/67 applied to territories extending beyond the common market; the Commission interpreted the exclusion in art. 3(b) fairly narrowly in *Distillers (Whisky and Gin)* [1985] OJ L369/19, [1986] 2 CMLR 664, CMR 10,750, in treating gin and whisky as not competitive. Moreover, in *Junghans GmbH (Re the Agreement of Gebrüder)* (77/100/EEC), [1977] OJ L30/10, [1977] 1 CMLR D82, CMR 9912 it treated the exemption narrowly, in that it did not apply where there was more than one dealer, although no others were to be supplied in the territory.

reciprocal agreement. 'Products' is used in the plural, so it may be that the regulation is excluded only if the parties compete in relation to both products.

The exclusion of the exemption to licences with joint ventures applies only when the parties are competing undertakings, and the licensing agreements relate to the activities of the joint venture. It is not clear whether it is excluded if the parties are not actual but only potential competitors.[32] On a literal construction, the exclusion may apply even if the joint venture has nothing to do with the products for which the parents compete. If A and B carry out several activities and compete in relation to Alpha, but have a joint venture to produce Beta, which the parents want for quite different and non-competing activities, the exclusion may apply. I cannot see why, unless the Commission is concerned that the habit of co-operation might extend from Betas to Alphas. It would have been better had the regulation been drafted to exclude the group exemption only if the parties competed in relation to the activities of the joint venture.

The difference between the treatment of joint ventures and reciprocal rights may seem perverse as the Commission might have been expected to favour joint ventures where resources may be integrated over reciprocal rights which may not be ancillary to such integration.

The most satisfactory solution would be for the first paragraph to be limited by the next two. This would leave the first paragraph to cover pools created through a joint organization used to acquire rights from outsiders, whether or not the members compete. This would, however, be difficult to reconcile with the idea that the regulation does not apply where there are more than two parties.

It may still be difficult to distinguish some joint ventures from other reciprocal exclusive rights. It would be sensible to treat as joint ventures only arrangements under which benefits of integration can be achieved.

5.1.2 Pure Sales Licences (Recital 8 and Article 5(1)(5))

Recital 8 states that since the object of the regulation is to facilitate the dissemination of technology and the improvement of manufacturing processes, it should apply only where the licensee manufactures the licensed products himself or has them made for his account.[33] Similar rules apply whether the product is goods or services. Presumably the protection

---

[32] Only the words 'competing manufacturers' and 'manufacturers of competing products' are defined in art. 10(17). It may be argued that 'competing undertakings' is a wider term.

[33] This exclusion was made in the know-how reg. at the suggestion of the European Parliament. See the explanatory statement to the motion for a resolution of the Committee on Economic and Monetary Affairs and Industrial Policy , 23–24 Mar. 1988, EP Doc. A2–36/88, [1988] 4 CMLR 653, 659.

from passive sales for five years that has always been permitted under the technology transfer regulations should not be permitted as between dealers.[34]

Article 5(1)(5) implements the recital simply by providing that the regulation shall not apply to 'agreements entered into solely for the purpose of sale'. Unlike recital 5 of the know-how regulation giving effect to *Delta Chemie/DDD*,[35] there is no provision that the technology transfer regulation applies when the licensor manufactures for a preliminary period before the licensee himself commences production.

The question arises whether the precedent of the Commission in *Delta Chemie* will be followed. The case was decided in relation to the patent licensing regulation which contained no provision about goods bought in for an initial period. In my view, it was always misconceived, as protection from passive sales was permitted when the licensee was operating only as a distributor, and the period of protection from passive sales permitted might well expire before manufacture commenced and the licensee might need the incentive to set up the plant. Moreover, the Commission's view could be used as an avoidance device to enable the holder to restrain his exclusive distributors from passive sales outside the territory. It might not be easy to establish that the parties never intended the distributor to start production from manufacture by other licensees in the licensed territory. Nevertheless, this may be a matter on which the new regulation is narrower in scope than the know-how regulation.

Recital 8 does not imply, as did article 1(5) of the know-how regulation, that provisions in the regulation other than article 1(1), sub-paragraphs (2), (3), (5), and (6) might apply, permitting a territorial restraint of production. This may be another example when the new regulation is of a narrower ambit than the know-how regulation.

5.1.2.1 The Need to Exclude Sales Licences from the Exemption

In the patent licensing regulation it was necessary for the Commission expressly to exclude dealers who might be given a sales licence under the patent and claim the exemption in the same way as a manufacturing licensee. The holder of a patent can restrain sale, so a distributor of a product protected by patent can be described as a licensee. That is true in relation to know-how licences when at least part of the technology is protected by patents. If pure sales licensees are not to be allowed to accept

---

[34] Reg. 1983/83, which grants a group exemption for exclusive distribution agreements permits a restriction on actively seeking customers outside the dealer's territory indefinitely, but no restriction on passively accepting unsolicited offers.

[35] [1988] OJ L309/34, [1989] 4 CMLR 535, 1 CEC 2254, described at 3.1.5 above.

a restriction on passive sales, therefore, recital 8 and article 5(1)(5) were needed.

It is difficult under this regulation to envisage a licensee of pure know-how which does not itself manufacture or have the products made for it. A dealer that does not make the product or have it made would not need to receive the substantial and secret know-how that is necessary to bring the agreement within the regulation. The exclusion of pure sales licences is not important for pure know-how agreements.

5.1.2.2 Manufacture for the Licensee's Account

Recital 8 expressly provides that a licensee may have the goods manufactured for his account without losing the benefit of the group exemption. Community law is concerned with undertakings rather than legal persons. So companies in the same corporate group following the policy of their parent are treated as one,[36] whether or not they are 'connected undertakings' as defined in article 10(14). There is no need to exempt agreements or instructions between them. One subsidiary may produce the goods for other subsidiaries of the same parent, and a restriction on manufacture by the latter would not infringe article 85(1) provided that the subsidiaries are following the policies of their parent company.

A licensee who has the product made by a sub-contractor[37] may well accept the risk of tooling up. He may provide the key tools or dies to the sub-contractor as well as the technology, and finance the operation. The sub-contracting notice implies that a sub-contractor is so closely integrated into the contractor's organization that it should not be treated as a separate undertaking. The justification applies even more strongly when a connected company produces. The technology regulation applies only to the agreement between licensor and licensee, but a contract with the sub-contractor needs no exemption.

5.1.2.3 What Amounts to Manufacture?

The question arises of what amounts to manufacture. Recital 7 of the regulation exempting patent licences stated that sales licences are governed

---

[36] See *Centrafarm BV* v. *Sterling Drug Inc.* (15/74), [1974] ECR 1147, [1974] 2 CMLR 480, CMR 8246, and *Viho Europe BV* v. *Commission* (T–102/92), [1995] ECR II–217, [1995] 4 CMLR 299, [1995] 1 CEC 562, appeal filed (73/95P). Lenz AG recommended on grounds of policy that collusion with a wholly-owned subsidiary falls outside art. 85(1). The Court's judgment is awaited. See 5.2.3 below.

[37] See the sub-contracting notice of the Commission [1979] OJ C1/2.

by regulation 1983/83.[38] This grants a group exemption for exclusive distribution agreements. Consequently, under the patent regulation it could be argued that any processing that would take the agreement outside regulation 1983/83 amounted to manufacture and sufficed to bring the agreement within the qualification of article 1(2) of the patent regulation. Consequently, a dealer who does too much processing to come within the exemption for distribution could be restrained under the patent regulation for five years from making passive sales.

There is no mention of regulation 1983/83 in the recitals to the know-how or technology transfer regulations, and the position of a licensee who does too much processing to come within regulation 1983/83 is not clear. Article 5(1)(5) refers to 'agreements solely for the purpose of sale'; recital 8 to the converse concept: 'manufacture by the licensee or for his account'. It can be argued that manufacturing in article 5(1)(5) should be interpreted to include any processing that takes the agreement outside the concept of 'resale' and one may refer to any case law under regulation 1983/83 if it develops.[39] It would be unfortunate if a gap had been opened between the two regulations.

The converse problem arises from the considerable amount of processing allowed under regulation 1983/83 according to the Commission's guidelines.[40] Can a dealer that performs considerable processing but whose agreement falls within regulation 1983/83 accept a restriction on passive sales under the technology transfer regulation? There may be some overlap between the two regulations.

5.1.2.4 Place of Manufacture

It does not matter where the licensee manufactures or has the products made, as long as it is not just a dealer. I see no reason why he should not be allowed to manufacture or have the products made outside his sales territory, even outside the Common Market or EEA.[41] This should enable

---

[38] Analysed by **Valentine Korah** and **Warwick A. Rothnie**, *Exclusive Distribution and the EEC Competition Rules—Regulations 1983/83 and 1984/83* (2nd edn., London: ESC Publishing, 1992), 80.

[39] Apart from *Moosehead/Whitbread* [1990] OJ L100/32, [1991] 4 CMLR 391, [1990] 1 CEC 2127, there has been no case law on licensing since the know-how reg. was adopted, and little on the interpretation of reg. 1983/83 since its adoption.

[40] Some guidance is given in the Commission's notice on the meaning of 'for resale' in regs. 1983/83 and 1984/83, [1994] OJ C101/2, at para. 9. It refers to processing that does not change the identity of the goods, nor add much value to what was sold to the dealer. This notice, however, binds no one save, perhaps, the Commission. See **Korah** and **Rothnie**, n. 38 above, at 11, n. 15.

[41] 'Exploitation' is defined in art. 10(10) as 'any use of the licensed technology in particular in the production, active or passive sales in a territory even if not coupled with manufacture in that territory, or leasing of the licensed products'.

licensees to obtain any available economies of scale or scope in production. It may also have the products made for it where labour is cheaper than it is in Europe, for instance in Taiwan. It seems from *Delta Chemie*, however, that it may not be able to have the goods made for it by the licensor, save for an initial period. It is not clear to me whether the precedent would apply to an arrangement to have another licensee make the products for an initial period.

In *Windsurfing International* v. *Commission*,[42] the Court objected to a provision requiring the licensee to manufacture only in a specific factory in Germany. The Court said that this was not justified on the ground that the patentee wanted to be able to control quality, and did not come within the specific subject matter of the patent. It did not say why the clause was anti-competitive.

5.1.2.5 Examples when the Technology Holder's Network Includes Exclusive Licensees for Some Territories and Exclusive Distributors for Others

Where the holder of technology exploits it by producing enough to supply only part of the requirements of the common market and arranges for this to be brought to market by exclusive dealers in some Member States, but licenses manufacture in others, recital 8 does not analyse which firms can be protected under the exclusive distribution regulation and which under the technology transfer one.

On a literal interpretation of the regulations, an exclusive distributor can be required under regulation 1983/83[43] not to make active sales outside its territory for an indefinite period. An exclusive know-how licensee, however, can be required not to manufacture or make active sales in the territory of other licensees for ten years, or passive sales for five, from the time when the goods were first put on the market within the common market by a licensee.[44] The focus is anomalous. The regulations should refer not to the undertakings to be restrained, but those being protected.[45]

Suppose that the holder of the technology, some of which is protected by

---

[42] *Windsurfing International* [1983] OJ L229/1, [1984] 1 CMLR 1, CMR 10,515, on appeal, *Windsurfing International Inc.* v. *Commission* (193/83), [1986] ECR 611, [1986] 3 CMLR 489, CMR 14,271, paras. 82–8.

The judgment was signed by only three judges, owing to the illness of Judge Joliet, and could easily be overruled by a full Court.

[43] Art. 2(2)(c).

[44] Art. 1(1)(4)–(6) and (2)–(4). See 6.6.5—6.7 below.

[45] **René Joliet** 'Trademark Licensing Agreements under the EEC Law of Competition' (1983) 5 *Northwest J Int'l. L & Bus.* 755, 802, observes that exclusive licences and the ancillary export bans raise the same issues of policy.

patent and includes a substantial element of secret and identified know-how, H, manufactures in England and supplies the United Kingdom itself. H grants an exclusive licence to G to make and sell in Germany and to F to sell in France. Suppose that F is required to buy in the product from G and sells it in France.[46] The licence to F is a pure sales licence and article 5(1)(5) clearly applies. F can, presumably, be restrained indefinitely by G or H under regulation 1983/83, which exempts exclusive distribution agreements,[47] from actively selling outside France but not from passive sales anywhere.

Since H has granted F a sales licence, on a literal construction of article 1(1), F may qualify as a 'licensee' of the patented technology and be protected under article 1(1)(4–6) of the technology transfer regulation from manufacture within its territory and direct sales there by G. I do not mean that F's licence would be exempt under the regulation—that is prevented by article 5(1)(5)—but that F would qualify as an 'other licensee' within the meaning of article 1(1)(4)–(6), so could be protected by G's licensing agreement.[48]

G can be restrained under article 1(1)(4)—1(1)(6) of the technology transfer regulation from making even passive sales into the territories of other licensees such as F for five years from the first time the protected goods were put on the market in the common market by a licensee and from manufacture and active sales for ten or for as long as there is a patent in both Germany and France, whichever be longer. G can also be restrained from making active or passive sales in the United Kingdom and Ireland for ten years from the same date or as long as there is a patent in both Germany and the United Kingdom. If France was reserved as 'the territory of the licensor' within the definition of article 10(12), then G can also be restrained from selling passively into France for ten years.

If F has no exclusive territory, no group exemption[49] will apply to the contract between it and H, and an individual exemption may be needed to protect the territories of G and H from active or passive sales by F.

If F has an exclusive sales licence, he can be kept out of G's territory to

---

[46] And that F has made no additions to the technology that he has passed on to G, so the sub-contracting notice does not apply as between F and G.

[47] It is due to expire at the end of 1997, but is likely to be renewed, possibly with alterations.

[48] This seems right on a literal reading. Nevertheless, the reg. is not at all clear about which licensee may accept a restriction and which may be protected. The Commission may well not have wanted a sales licensee to receive such protection, especially protection from passive sales.

If the territory was reserved to H in the licensing agreement with G, the problem does not arise, but there seems to be no reason why the result should depend on whether H granted F a sales licence.

[49] Reg. 1983/83 applies only if there is an exclusive territory. See *Junghans*, n. 31 above.

the extent allowed by regulation 1983/83: F could be required not actively to sell outside France, either in the United Kingdom or Germany, and H could agree not to sell in France to dealers or to consumers. H could require G not to sell actively or passively in France for ten years from the first sale by a licensee under the technology transfer regulation if H had reserved France for itself.[50] It is not clear that a promise by H to impose such a restriction on G would be exempt under the distribution or technology transfer regulation, although I would argue from the recitals to both regulations and from the reference in the preamble to the EC Treaty to 'fair competition' that it must be.[51] Some officials, educated in civil law countries, state that such a promise is the counterpart of the restrictions accepted by each licensee and so permitted.

The relative amounts of protection are ludicrous. They operate the wrong way round from a policy point of view even if one accepts that there should be detailed rules rather than an economic analysis of the amount of protection required to induce investment. It is G who has had to invest in tooling up and who might have refused to do so unless protected from passive sales by F, not *vice versa*. On grounds of policy, one would expect the requirement of manufacture to relate to the licensee to be protected, not the licensee restrained. On a literal view, however, it seems to be the licensee accepting the restriction on exploiting elsewhere that must manufacture or have the product made for the agreement to be exempted under the technology transfer regulation. Similar objections apply to other group exemptions.

Suppose that H licenses F to manufacture as well as sell in France, but that F decides to buy in the product from G for an initial period while it creates sufficient demand to warrant tooling up. Would that be treated as a technology licence for the initial period on the basis of *Delta Chemie*?[52] There is no reason of policy to distinguish such a case from initially buying in from the licensor. There is no express provision about initially buying the product from the licensor or another licensee in the technology transfer regulation or in the patent licensing regulation, on which *Delta Chemie* was decided. Common sense dictates that the licensee must be given time to set up production, and it may be commercially sensible and pro-competitive to use the time to develop the market through distribution.

If H should also license N to manufacture in Norway and sell there, in Finland, and in Sweden, the EEA regulation will apply. N can be

---

[50] In that event, F would presumably buy from H rather than another licensee.

[51] This must apply also to the EEA. Great pains have been taken to harmonize the rules with those of the EEC and create a homogenous EEA, as required by art. 1 of the EEA Agreement. See 4.2.1 above.

[52] Cited at n. 35 above, and described at 3.1.5 above.

restrained from exploiting at all in the United Kingdom for ten years from first marketing in the EEA by a licensee or for as long as there remains a valid patent in both the United Kingdom and Norway or, possibly,[53] in the United Kingdom and in Sweden or Finland. N can be restrained from exploiting in Germany or France by manufacture or active selling for the same period[54] and from passive sales for five years. H could also reserve his right under article 2(1)(14)[55] indefinitely to use his patents in Germany etc. to restrain direct sales by N there, but could not restrain a purchaser from N selling in Germany because of the doctrine of exhaustion which applies to products of EEA origin.

If H should sell to S in Switzerland or grant him a manufacturing licence, H could use his patent rights in common market countries to restrain direct imports from Switzerland and his Swiss patent to restrain imports into Switzerland, direct or indirect. I would argue from the benefits the Commission recites are to be expected from exclusive licences in recital 12 that H may promise any licensees in the EEA to exercise this right.

5.1.3 Franchising Agreements (Recital 8 and Article 5(1)(4))

Recital 8 continues:

> Also to be excluded from the scope of this Regulation are agreements relating to marketing know-how communicated in the context of franchising arrangements.

As in the case of pure sale licences, the technology transfer regulation does not recite that exclusive franchising agreements may be exempt under the group exemption for exclusive distribution and service franchises, regulation 4087/88. If an agreement is exempted by the group exemption for franchising it does not matter whether or not it be exempted by that for technology transfer, but under the franchising regulation, as under the group exemption for exclusive distribution, there is no exemption for any restraint on passive sales, and franchisees must be permitted to sell to each other anywhere. A franchisor might want to encourage investment in developing a market for tooling-up by granting protection from passive sales, and he might want to discriminate in price between Member States, in which case he would have to prevent cross-sales between franchisees. The use of the technology transfer regulation for this purpose is excluded pursuant to recital 8.

---

[53] The reg. does not address the problem of the duration of the territorial restraints where there is a valid patent only in parts of the licensed territory. In so far as there is qualifying know-how, the periods of years prescribed in art. 1(3) and (4) must apply, but the period based on the life of the patent in both the territory licensed and that to be protected is unclear when the patent is valid only for part of one or both of the territories. See 6.6.5 below.

[54] It is the life of the German rather than the UK patent that would be relevant.

[55] Analysed at 7.3.14 below.

A technology licensee is not likely to develop a sales market by becoming a franchisee before beginning to manufacture. So the possible application of *Delta Chemie*[56] is unlikely to matter.

The exclusion from the technology transfer regulation is most serious for what a lawyer might call a trade mark licence and his client an industrial franchise—instructions on how to produce something, and the crucial licence of a trade mark. There is no group exemption available for these unless the technology transfer regulation can apply: see 5.1.4 below.

5.1.4 Provisions Relating to Marks and Other Industrial Property Rights (Recital 6 and Article 5(1)(4))

By virtue of article 8 various rights, such as plant breeders' rights, are deemed to be patents for the purposes of this regulation. These are considered at 5.2.4 below. This section is concerned with intellectual property rights other than patents of a kind not listed in article 8 such as trade marks, design, and copyright.

Agreements relating to any kinds of industrial property rights may be the subject of a block exemption under regulation 19/65, article 1(1)(b), as is stated in the first recital of the technology transfer regulation.

Article 1(1) of regulation 240/96 extends the scope of the group exemption from patent and/or know-how licences to those, 'containing ancillary provisions relating to intellectual property rights other than patents'.

Recital 6 states that:

> It is appropriate to extend the scope of this Regulation to pure or mixed agreements containing the licensing of intellectual property rights other than patents (in particular, trademarks, design rights and copyright, especially software protection), when such additional licensing contributes to the achievement of the objects of the licensed technology and contains only ancillary provisions.

Article 5(1)(4) implements the final qualification and provides that the regulation shall not apply to:

> licensing agreements containing provisions relating to intellectual property rights other than patents which are not ancillary;

The French text, in which the technology transfer regulation was drafted, ends *qui ne sont pas des clauses accessoires*, which makes it clear that it is the provisions rather than the licences that must be ancillary.

For the regulation to apply, first, the provisions relating to other

---

[56] Cited n. 35 and discussed at 3.1.5 above.

intellectual property rights must be ancillary;[57] secondly the additional licensing must contribute to the objects of the licensed technology; and thirdly, any restrictions of competition attached to those rights must be also attached to the technology licence and exempted by the regulation.

Article 10(15) defines 'ancillary provisions' as:

> provisions relating to the exploitation of intellectual property rights other than patents, which contain no obligations restrictive of competition other than those also attached to the licensed know-how or patents and exempted under this Regulation;

This is a surprising definition, since it does not require the provision to be ancillary in the usual sense that it be of less value than the technology licence. These provisions are different from those in article 5(1)(4) of the know-how regulation. Under that regulation, it was the intellectual property rights rather than the provisions relating to them that had to be ancillary for the agreement to qualify for the exemption.

The Commission decided in *Moosehead/Whitbread*[58] that the know-how regulation did not apply to an industrial franchise where the trade mark licence was crucial. Although the Commission noted at paragraph 15(4)(b) that the Moosehead 'trade mark is comparatively new to the lager market in the territory', it concluded at paragraph 16(1) that regulation 556/89 did not apply because the parties considered the Canadian origin of the mark was crucial to the success of the marketing campaign. The mark was, therefore, not 'ancillary' as required by articles 1(1) and 5(1)(4) and the know-how regulation did not apply.

It is not entirely clear to me how article 5(1)(4) of the new regulation should be construed. If one substitutes the definition for the words 'ancillary provisions', there is no need for the provisions relating to other kinds of intellectual property to be less valuable than the technical know-how: it suffices that there be no additional restrictions of competition.

This literal reading may, however, be inconsistent with the penultimate provision of recital 6, which extends the regulation to ancillary provisions 'when such additional licensing contributes to the objects of the licensed technology'.

This may imply that the patent or know-how, and not just the provisions, should be more important than the other intellectual property rights licensed. Nevertheless, I would argue that in *Moosehead/Whitbread* the mark contributed to the objectives of the licensed know-how by helping

---

[57] The Economic and Social Committee objected to this concept when commenting on the know-how reg. See the Additional Opinion of the Economic and Social Committee, adopted at the 254th plenary session on 23 Mar. 1988 [1988] OJ C134/10, [1988] 4 CMLR 498.

[58] N. 39 above; analysed at 3.1.7 above.

the franchisee to develop demand for the final product. The decision rejected the application of the group exemption on the ground that it was not ancillary, not because it did not 'contribute to the objects of the licensed technology'.

The narrow interpretation of the qualification 'ancillary' is somewhat supported by the wording of recital 8:

> 'Also to be excluded from the scope of this Regulation are agreements relating to *marketing* know-how communicated in the context of franchising arrangements' [my emphasis].

It is clear that the mere grant of a trade mark licence does not prevent the application of the regulation. Article 1(1)(7) expressly exempts:

> an obligation on the licensee to use only the licensor's trademark or get up to distinguish the licensed product during the term of the agreement, provided that the licensee is not prevented from identifying himself as the manufacturer of the licensed products;

In industrial franchising there is usually technical rather than, or as well as, marketing know-how. I understand that officials take the view that industrial franchises may be brought within the regulation as long as there is sufficient qualifying know-how or a patent licensed.

Sometimes it will be necessary to insert provisions restrictive of competition relating to distribution that are not expressly exempted by the technology transfer regulation, but if the regulation does apply to industrial franchises one might notify the agreement under the opposition procedure,[59] and the Commission could validate these provisions and the exclusive territory by not opposing the application of the regulation. It would not have to adopt a formal decision of exemption. After four months, a binding position might be obtained.

The words 'ancillary provisions' are not meaningless, even if they do not refer to the other intellectual property rights being of lesser value. One cannot avoid the effects of the black list or opposition procedure by attaching to the copyright, trade mark, or other intellectual property rights provisions that would be subject to article 3. This was to be expected. What is new is the absence of any idea that the provisions relating to other rights should be of less value than the patents or know-how.

It is arguable, however, that the right to use a mark can never be ancillary to a franchise: the mark is of the essence. Professeur Joliet suggested shortly before he joined the Court[60] that franchising and trade mark licences should be distinguished on the ground that production franchising is usually local. He instanced the services of restaurants, hotels,

---

[59] Described in Ch. 8 below.   [60] **Joliet**, n. 45 above.

dance schools, and rent-a-car companies. Under the Commission's current terminology these would be called 'service franchises' and may be exempted by the franchising group exemption if an exclusive territory infringes article 85(1). That regulation does not cover the kind of production franchises that Joliet would characterize as trade mark licences, such as that in *Moosehead/Whitbread,* which the Commission found was not covered by the know-how regulation. That franchise was not local—it covered the whole of the British Isles.

In *Pronuptia,*[61] the Court implied that the group exemption for exclusive dealing does not apply to distribution franchising, because, *inter alia*, of the provisions relating to the mark, although it had just held that they did not restrict competition: franchising is different from distribution, and may be from a know-how licence. Those educated in the civil law tradition tend to be more formalistic than common lawyers and many consider that there is a proper category for each kind of transaction.[62]

As stated at 3.1.2 above, Campari has now changed its network and makes the bitters itself for sale throughout the common market, using exclusive distributors and subsidiaries rather than licensees.[63] If it or other firms are induced to do this in order to come within the group exemption for exclusive distribution, the Commission's separate codes for different transactions must have increased inefficiency at least at the margin.

James Venit suggests that this is because distribution agreements are recognized to be vertical.[64] The technology regulations have not been drafted to exclude all horizontal agreements. The provisions of article 5(1)(1)–(3) prevent the application of the regulation to some horizontal agreements,[65] but not to all.

The argument about industrial franchises applies also to licences of intellectual property rights other than marks, such as designs or copyright. If the trade marks, design rights, or software do not have to be ancillary, not only may industrial franchise agreements but also licenses of other intellectual property rights, in particular software, be brought within the regulation, provided there is a relevant patent or qualifying know-how licence as well.

---

[61] *Pronuptia de Paris GmbH* v. *Pronuptia de Paris Irmgard Schillgalis* (161/84), [1986] ECR 353, [1986] 1 CMLR 414, CMR 14,245, para. 33.

[62] This may be derived from the Roman law of real contracts.

[63] *Eighteenth Report on Competition Policy* (Brussels: EC Commission, 1989), point 69.

[64] 'In the wake of *Windsurfing*: Patent Licensing in the Common Market', in **Barry Hawk** (ed.), *Thirteenth Annual Proceedings of the Fordham Corporate Law Institute* (New York: 1986) 517, at 528–9.

[65] Analysed at 5.1.1–5.1.1.3 above. Art. 5(1)(1) also prevents the exemption from applying to patent pools that are not between competitors and may be vertical.

5.2 INCLUSIONS (ARTICLE 6)

Article 6, like article 11 of the patent regulation, brings in sub-licences, assignments for royalty, or licensing agreements where the rights or obligations are assumed by connected undertakings.

5.2.1 Sub-licences (Article 6(1))

Article 6 provides that:

> This regulation shall also apply to:
> (1) agreements where the licensor is not the holder of the know-how or the patentee, but is authorized by the holder or the patentee to grant a licence;

The Commission is accustomed to several layers of distribution agreements, each exempted by regulation 1983/83, provided that there are only two parties to each. A brand owner may appoint an exclusive distributor for, say, France. Then that distributor may agree to sell only to a particular wholesaler for the Paris region, who supplies retailers, possibly also on an exclusive basis.

It must have seemed natural to extend the same concept to technology licences. Licensing seldom goes down to so many levels, but where sub-licensing is convenient commercially, there is no reason to exclude it from the exemption. It might have been better had the empowering regulation applied to vertical agreements rather than those to which only two undertakings are party.

Often, a licensor will want to control carefully any sub-licences granted in order to preserve the confidentiality of its know-how and, where a trade mark licence is granted, to ensure quality control.[66] A licensor may well prescribe the form of agreement for a licensee to use with any sub-licensee; but the exemption applies only when there are no more than two parties to the agreement, and often sub-licences will be tripartite, especially where a sub-licensee is authorized to use the trade mark of the technology holder and must accept that any reputation he makes for the mark will inure to the holder. Article 6(1) is most likely to be used when the licensor virtually assigns his know-how by granting an exclusive licence for the world to the sub-licensor and ceases to be concerned about its reputation or becoming publicly known.

Somewhat greater territorial protection from passive sales may, possibly,

---

[66] The Court was not very sympathetic to this consideration in *Windsurfing International* v. *Commission*, n. 42 above, paras. 45–50.

be obtained by taking advantage of sub-licensing. Where a licence includes patented technology and/or qualifying know-how, licensor and licensee are allowed to agree to keep right out of each other's territory for ten years from the date when the licensed products were first put on the market within the common market by one of the licensees,[67] whereas one licensee may be restrained from passive sales in the territory of another for only five years from that date.

Instead of licensing L1 and L2 for different territories, the licensor, H, might license L1 for both territories, and L1 could then sub-license L2 for one of them. L1 and L2 might then accept restraints on passive sales into the other's territory for ten years from the first sale by a licensee in the common market.

Where the licensor has a network of other licences, this may not help much, unless transport costs are important, as the protection from passive sales by the other licensees will be limited to five years. It might be useful, however, if there are only two licensees in the common market, or if the marginal costs of one licensee are particularly low.[68] It might be given a sole and exclusive licence for a large territory and allowed to sub-license others if the problem of confidentiality can be overcome and there is no multipartite agreement or concerted practice embracing other licensees.

There may be a problem, however, as between H and L2. H may agree with L1 to keep out of both territories, but L2 might not be able to enforce this contract to which he is not a party. L2 may, however, not be permitted to agree not to sell in the territories expressly reserved for H, in which case it may be that H cannot obtain protection under article 1(1)(3). Moreover, unless the territory is licensed to another licensee, territorial protection under sub-paragraphs (4)–(6) is not available.

L2 can be kept out of territories reserved to other licensees of H, such as L3. Article 1(1)(4)–(6) is expressed in the passive voice—'territories licensed to other licensees'. It is not stated that the licence must be directly from L1. So probably L2 and L3 may be restrained in the terms of those sub-paragraphs.

This suggestion is contrived, but may be useful in exceptional agreements, although not as useful as under the know-how regulation, when the date from which the permitted territorial protection started differed for the different sub-paragraphs of article 1(1).

---

[67] Art. 1(1)(1)–(3).

[68] Some holders of technology arrange for the first licensee to perfect the technology and then charge them a lower royalty. Such a licensee might be told that it may pay a low royalty on its own production, but a higher one on that of its sub-licensees. It can then sub-license under separate bilateral agreements.

5.2.2 Assignments (Article 6(2))

The regulation also applies to:

> (2) assignments of know-how, patents or both where the risk associated with exploitation remains with the assignor, in particular where the sum payable in consideration of the assignment is dependent on the turnover obtained by the assignee in respect of products made using the know-how or the patents, the quantity of such products manufactured or the number of operations carried out employing the know-how or the patents;

This provision, which is also derived from article 6 of the know-how regulation, aggregates the provisions in recital 6 and article 11(2) of the patent regulation. Assignments for a lump sum seem to be excluded from the regulation and may be caught by article 85(1).[69] Royalties need not be proportionate to turnover, provided that they depend on it. Sometimes the proportionate royalty may fall off as production expands. The higher initial royalty may enable the parties to share the risk of the innovation not being a sufficient success.

5.2.3 'Connected Undertakings' (Article 6(3))

The regulation also applies to:

> (3) licensing agreements in which rights or obligations of the licensor or the licensee are assumed by undertakings connected with them.

'Connected undertakings' is defined in article 10(14). The definition attempts to give greater precision to the concept of 'undertaking' than the

---

[69] As early as *Sirena Srl* v. *Eda Srl and Others,* (40/70), [1971] ECR 69, [1971] CMLR 260, CMR 8101, the Court considered that the exercise of trade mark rights to restrain the import of goods bearing the same trade mark might infringe art. 85(1) when the trade mark for different territories had been assigned to different undertakings by an agreement that continued to have anti-competitive effects. The Court stated that:

> 'A trade-mark right, as a legal entity, does not in itself possess those elements of contract or concerted practice referred to in Article 85(1). Nevertheless, the exercise of that right might fall within the ambit of the prohibitions contained in the Treaty each time it manifests itself as the subject, the means or the result of a restrictive practice.'

These words have been repeated in many subsequent cases.

Nevertheless, an economic assessment would have to be made in the legal and economic context of the assignment to establish an infringement: *Ideal-Standard—IHT Internazionale Heiztechnik GmbH* v. *Ideal-Standard GmbH* [1994] ECR I–2789, [1994] 3 CMLR 857, [1994] 2 CEC 222.

It is difficult to see how the assignment of know-how not protected by any intellectual property rights could infringe art. 85(1) unless the assignment is to a competitor with competing technology in a concentrated market. In that case, the exemption might not to apply anyway if the assignment formed part of a technology pool or reciprocal exclusive rights, excluded by art. 5(1)(1) or (3).

Court has yet achieved. Often one company within a group will hold the patents, and license them and other technology to other firms in the same group. Such an agreement does not infringe article 85(1) if the legal persons involved are under common control.[70] Where the subsidiary holding the rights, however, grants a licence to an outsider, or where the subsidiary that has been granted a licence grants a sub-licence, the outsider is likely to need protection from other companies within the group if he is to commit himself to paying royalties and setting up production. The reasons for permitting territorial protection between the parties and between different licensees must apply to undertakings[71] rather than to legal persons.

Undertakings are 'connected' for the purposes of the regulation where a party to the agreement directly or indirectly owns more than half the shares, has power to exercise more than half the voting rights, and so on. Parents of a 50–50 joint venture are not connected with the joint venture or each other, but if one parent has more than half the voting rights and the other more than half the shares, each might be connected to the joint venture, but not to the other.

5.2.4 Rights to be Treated as Patent Rights (Article 8(1))

Article 8 repeats and extends article 10 of the patent regulation, which was omitted from the know-how regulation. It provides that various rights shall be deemed to be patents for the purposes of the regulation. There is little to be said about most of the items, but something will be said about the new items.

The patent regulation treated as patents the following items, which are also so treated by article 8 of the technology transfer regulation:

(a) patent applications;
(b) utility models;

---

[70] In *Centrafarm* v. *Sterling*, n. 36 above, the Court overruled its AG and stated that the allocation of tasks between companies subject to common control is outside art. 85(1).

In *Viho Europe BV* v. *Commission*, n. 36 above, the Court of First Instance confirmed the Commission's decision that export bans imposed by a parent company on its wholly-owned subsidiaries did not infringe art. 85(1). Judgment is awaited, but Lenz AG recommended on grounds of policy that art. 85(1) should not apply to collusion within the corporate group whether or not it related to the allocation of tasks.

See also *Hydrotherm*, n. 31 above, where Signor Andreoli, a company, and a partnership he controlled were treated as only a single undertaking within the meaning of reg. 67/67.

The Court has not had to consider collusion when the subsidiary has not been wholly owned. It has, however, attributed to a parent company the acts of a half-owned subsidiary, when the subsidiary had followed its parents' wishes—*Commercial Solvents—Istituto Chemioterapico Italiano SpA and Commercial Solvents Corp.* v. *Commission* (6 & 7/73), [1974] ECR 223, [1974] 1 CMLR 309, CMR 8209. Contrast *Gosme/Martell* [1991] OJ L185/23, [1992] 5 CMLR 586, [1991] 2 CEC 2110, paras. 30–2.

[71] The introductory words to art. 1(1) refer to not more than two undertakings.

(c) applications for registration of utility models;

(e) *certificats d'utilité* and *certificats d'addition* under French law;

(f) applications therefor; as well as

2. agreements for the exploitation of inventions for which an application is made within the period permitted by international conventions.[72]

The new items are:

(d) topographies of semiconductor products;[73]

(g) supplementary protection certificates for medicinal products or other products for which such supplementary protection certificates may be obtained;[74]

(h) plant breeders' certificates.[75]

Technically a supplementary protection certificate does not extend the duration of a pharmaceutical patent to compensate for the time lost when patented products are undergoing their safety checks and clinical trials, but creates a *sui generis* right coming into force on the expiry of the patent. There was, therefore, some doubt whether it counted as a patent under the earlier group exemptions. This was particularly serious as, once the information required for health purposes has been published, no secret know-how remains.

---

[72] e.g. Paris Convention for the Protection of Industrial Property of 20 Mar. 1883, as last revised at Stockholm on 14 July 1967, 828 UNTS 11851, 305 permits patents to be applied for within one year of the first application in a member state of that convention and the priority date of the later applications is that of the original one.

[73] See Council dir. 87/54 of 16 Dec. 1986 on the legal protection of topographies of semiconductor products [1987] OJ L24/36.

There are also various Council decisions extending the legal protection of such products to persons from various countries with which bilateral treaties have been negotiated.

The initiative for this kind of intellectual property lies in the US where the Semiconductor Chip Protection Act 1984 created a *sui generis* right for 10 years protecting the industry against unfair copying outside the US.

[74] At the moment they may be obtained under national legislation implementing Council dir. 1768/92 on the creation of a supplementary protection certificate for medicinal products [1992] OJ L182/1. Although in June 1996 the Council adopted a reg. creating similar rights for pesticides, Commission brief No WE/22/96, 13 June 1996, no such addition has been made to art. 8.

[75] Plant breeders' rights are protected in most Member States, and in parallel under Council reg. 2100/94, [1994] OJ L227/1, which allows plant variety rights to be granted throughout the Community.

Plant breeders' rights were expressly excluded from the benefit of the patent reg. by art. 5(1)(4). The Commission considered that its decision in *Maize Seed* [1978] OJ L286/23, [1978] 3 CMLR 434, CMR 10,083, and the Court's judgment in *Nungesser (L. C.) KG and Kurt Eisele* v. *Commission* (258/78), [1982] ECR 2015, [1983] 1 CMLR 278, CMR 8805, had not provided sufficient experience.

Since then, the Court gave judgment in *Erauw-Jacquéry (Louis) Sprl* v. *La Hesbignonne Société Coopérative* (27/87), [1988] ECR 1919, [1988] 4 CMLR 576, [1989] 2 CEC 637, followed by the Commission in two informal decisions, *Standard Seed Production and Sales Agency in France*, [1990] OJ C6/3, [1990] 4 CMLR 259; and *Re the Application of CBA* [1995] OJ C211/11, [1995] 5 CMLR 730. See also *Royon* v. *Meilland*, sometimes known as *Plant Breeders' Rights (Roses)* [1985] OJ L369/9, [1988] 4 CMLR 193, CMR 10,757.

Article 8(3) also provides:

> (3) This Regulation shall furthermore apply to pure patent or know-how licensing agreements or to mixed agreements whose initial duration is automatically prolonged by the inclusion of any new improvements, whether patented or not, communicated by the licensor, provided that the licensee has the right to refuse such improvements or each party has the right to terminate the agreement at the expiry of the initial term of an agreement and at least every three years thereafter.

This provision is almost the mirror image of article 3(2) of the patent regulation and article 3(1) of the know-how regulation, which blacklisted agreements where the duration could be prolonged without each party being given the right to terminate annually.[76] It is now made clear that the regulation can apply when the condition in the proviso is fulfilled.

---

[76] Art. 3(2) of the patent reg. (2349/84 [1984] OJ L219/15) provided that the exemption should not apply where:

> 'the duration of the licensing agreement is automatically prolonged beyond the expiry of the licensed patents existing at the time the agreement was entered into by the inclusion in it of any new patent obtained by the licensor, unless the agreement provides each party with the right to terminate the agreement at least annually after the expiry of the licensed patents existing at the time the agreement was entered into, without prejudice to the right of the licensor to charge royalties for the full period during which the licensee continues to use know-how communicated by the licensor which has not entered into the public domain, even if that period exceeds the life of the patents;'

The provision was explained in recital 20, which added that the parties are free to enter into new agreements concerning new patents.

Art. 3(10) of the know-how reg. was identical, save that the provision about royalties was expressed in recital 15.

6

The Exemption (Article 1)

The exemption provided by the technology transfer regulation is broader and simpler than that under the earlier group exemption. The drafting has been improved in many ways. There are few agreements that would qualify under the earlier regulations and do not qualify under this one. For a list see 9.2.1 below.

6.1 INTRODUCTORY WORDS AND DEFINITIONS

Article 1 follows the structure of the earlier regulations exempting licensing agreements, and exempts pure patent, pure know-how, and mixed patent and know-how licences between no more than two undertakings that include one or more of the obligations listed in article 1(1). The introductory words provide:

> 1. Pursuant to Article 85(3) of the Treaty and subject to the conditions set out below, it is hereby declared that Article 85(1) of the Treaty shall not apply to pure patent licensing or know-how licensing agreements and to mixed patent and know-how licensing agreements and to mixed patent and know-how licensing agreements, including those agreements containing ancillary provisions relating to intellectual property rights other than patents, to which only two undertakings are party and which include one or more of the following obligations:

Six of the eight following obligations relate to a sole territory and various restrictions on exploiting the technology in other territories. Article 1(5) provides that obligations of the same types but of a more limited scope are also exempt.

Most of the terms in the introductory words are defined in article 10, but some are described in the recitals, as they were in the patent regulation. 'Licensing agreement' is defined in article 10(6) simply as:

> pure patent licensing agreements and pure know-how licensing agreements as well as mixed patent and know-how licensing agreements;

There is no requirement that the know-how should be ancillary to the patent as in the patent regulation, or *vice versa*. This is most welcome. The problem of deciding whether the know-how was ancillary and covered by the patent regulation was artificial. Normally both the patented and unpatented know-how are required to achieve an efficient result. Consequently, there were problems in forecasting whether the Commission or a court would consider that the patent regulation applied.

Recital 5 states that:

> (5) Patent or know-how licensing agreements are agreements whereby one undertaking which holds a patent or know-how ('the licensor') permits another undertaking ('the licensee') to exploit the patent thereby licensed, or communicates the know-how to it, in particular for purposes of manufacture, use, or putting on the market. In the light of experience acquired so far, it is possible to define a category of licensing agreements covering all or part of the common market which are capable of falling within the scope of Article 85(1) but which can normally be regarded as satisfying the conditions laid down in Article 85(3), where patents are necessary for the achievement of the objects of the licensed technology by a mixed agreement or where know-how—whether it is ancillary to patents or independent of them—is secret, substantial and identified in any appropriate form. These criteria are intended only to ensure that the licensing of the know-how or the grant of the patent licence justifies a block exemption of obligations restricting competition. This is without prejudice to the right of the parties to include in the contract provisions regarding other obligations, such as the obligation to pay royalties, even if the block exemption no longer applies.

Recital 4 defines some of these terms:

> This Regulation should apply to the licensing of Member States' own patents, Community patents and European patents ('pure' patent licensing agreements). It should also apply to agreements for the licensing of non-patented technical information such as descriptions of manufacturing processes, recipes, formulae, designs or drawings, commonly termed 'know-how' ('pure' know-how licensing agreements), and to combined patent and know-how licensing agreements ('mixed' agreements), which are playing an increasingly important role in the transfer of technology. For the purposes of this Regulation, a number of terms are defined in Article 10.

There is one problem that might theoretically arise over a licence for a non-member state under a patent not granted under the European Patent Convention where there is no additional know-how. Such a licence would be granted under a patent of a kind that is not mentioned in recital 4, yet according to recital 7, analysed at 4.2 above, if the licence has effects within the common market, perhaps through export bans of the scope exempted by article 1(1), it should be covered by this regulation. Such a case must be rare and it is thought that in view of the lack of exclusive wording in recital 4, recital 7 must prevail to exempt the agreement.

6.1.1 'Know-how' (Article 10(1))

Article 10 provides that:

> For purposes of this regulation:
> (1) 'know-how' means a body of technical information that is secret,
> substantial and identified in any appropriate form;

This is the same definition as in the know-how regulation. Recital 4, like recital 1 of the know-how regulation, instances 'descriptions of manufacturing processes, recipes, formulae, designs or drawings'. It should be contrasted with 'marketing know-how communicated in the context of franchising arrangements' licences of which are excluded from the regulation by recital 8 and article 5(1)(4), as discussed at 5.1.3 above.

During the consultations on the know-how regulation, some of the Member States wanted the regulation to be restricted to substantial, secret know-how to prevent competitors arranging to allocate markets under guise of a licence of trivial know-how.[1] In my view, this was misconceived, because if the know-how is trivial so is the protection from licensor and other licensees. The only provisions permitted by the regulation that do not relate to the technology licensed or products made thereby are the best endeavours clause now whitelisted in article 2(1)(17) and (18) and the provisions for minimum payment or a minimum number of operations provisions, allowed by article 2(1)(9), that may discourage the use of alternative technology.[2] Nevertheless, the Commission, by defining the words very broadly, weakened the requirements of secrecy and substantiality considerably over the period while its drafts of the know-how regulation were being discussed.

The final paragraph of article 1(3) of the new regulation provides that the exemption applies only so long as the know-how remains secret and substantial and provided that it is recorded.[3]

6.1.1.1 'Secret' (Article 10(2))

To satisfy the opinion of the European Parliament on the draft know-how

---

[1] **Sebastiano Guttuso**, in **Barry Hawk** (ed.), *Thirteenth Annual Proceedings of the Fordham Corporate Law Institute* (New York: 1986) 477, 491.

[2] The Commission has power to withdraw the exemption under art. 7(4) if the parties are competing manufacturers.

[3] Where patents are also licensed, recital 16 and art. 1(4) provide that the territorial obligations may last as long as there are patents in both the territory licensed and that to be protected.

Whether the exemption of art. 2(2) can outlast the necessary patents licensed or the secrecy of substantial know-how is unlikely to be important, given that recital 18 states that these obligations seldom restrict competition.

regulation,[4] the know-how must be secret in order to qualify. In article 10(2) of regulation 240/96, 'secret' is defined broadly and in relative terms so that little know-how for which it is worth paying a royalty will fail to qualify:

> (2) 'secret' means that the know-how package as a body or in the precise configuration and assembly of its components is not generally known or easily accessible, so that part of its value consists in the lead which the licensee gains when it is communicated to him; it is not limited to the narrow sense that each individual component of the know-how should be totally unknown or unobtainable outside the licensor's business;

As Jim Venit has observed,[5] this definition is broad enough to include the combining of known technical elements to produce a new configuration, but it is not clear whether it embraces a package of know-how that is known even in its totality, but is to be used for a different, hitherto unknown, purpose.

The broad definition is helpful, but some uncertainty must remain with terms such as 'generally known' or 'easily accessible'. To whom must it be known, the world in general, or those in the particular industry? In *Boussois/Interpane*,[6] the Commission said at the end of paragraph 2 that the know-how was:

> secret in the sense that, although individual components of it may not be totally unknown or unobtainable, *particularly for qualified engineers in the industry*, the know-how package as a whole is not readily available, and can therefore be said to be not in the public domain [my emphasis].

The test is pragmatic and relates to firms like the licensee's.

Problems may arise when the know-how is disclosed shortly after the first sale, because the product can be reverse-engineered. Territorial protection that ends shortly after the first sale of the product is of little value, even though the payment of royalties may clearly be extended.[7] Secrecy will not end on the first sale, however, but when it is established that sufficient people have done the reverse-engineering, so that it can be said to be 'generally known'. 'Easily accessible' may refer to what can easily be found in the obviously relevant books in a library, not in some doctoral thesis hidden in the library of the university that granted a degree, or an obscure footnote in a book dealing with something quite different. In less extreme cases it will not be possible to give clear advice on the meaning of these concepts.

---

[4] See also explanatory statement to the motion for a resolution of the Committee on Economic and Monetary Affairs and Industrial Policy , 30 Mar. 1988, EP Doc. A2–36/88 [1988] 4 CMLR 653, at 655.

[5] In notes prepared for a conference organized in London by ESC in Nov. 1988.

[6] *Boussois/Interpane* [1987] OJ L50/30, [1988] 4 CMLR 124, CMR 10,859.

[7] Recital 21.

The definition of 'secret' may well be wide enough for the qualification not to be a serious limitation. It is thought that disclosure to third parties bound by a duty of confidentiality does not prevent the information being secret. If the technology is disclosed only in isolated geographic or sectoral areas it will not be widely known.

6.1.1.2 'Substantial' (Article 10(3))

In article 10(3) 'substantial' is also defined broadly:

> (3) 'substantial' means that the know-how includes information which must be useful, i.e. can reasonably be expected at the date of conclusion of the agreement to be capable of improving the competitive position of the licensee, for example by helping him to enter a new market or giving him an advantage in competition with other manufacturers or providers of services who do not have access to the licensed secret know-how or other comparable secret know-how;

This definition is also broad enough to cover any know-how for which the licensee is likely to be prepared to pay. The definition includes know-how that gives the licensee some lead time to enter a new market or compete more effectively. The lead time may be shorter than the periods of territorial protection permitted: the agreement may qualify for the block exemption if it is 'useful . . . at the date of conclusion of the contract'. That is what the provision says and that is the time at which the licensee has to decide whether to incur sunk costs in tooling up and making a market. In *Rich Products/Jus-Rol*,[8] the Commission stated that:

> In view of these considerations, it is evident that the technical know-how is of particular value to Rich Products and is of great importance for the production of frozen yeast dough by Jus-Rol. Therefore, the know-how provided by Rich Products has to be considered as substantial and capable of improving the performance of the licensee who is thus willing to pay royalties for it.

If the final sentence represents the meaning of 'useful' in the regulation, most licences will qualify.[9]

---

[8] *Rich Products/ Jus-Rol* [1988] OJ L69/21, [1988] 4 CMLR 527, CMR 10,956, point 4, para. 3.

[9] It is only fair to add that the Commission had just referred to the difficulties of developing frozen yeast dough and the substantial research work done on it by Rich Products. It may also have been relevant that it added to the licensee's product range, and brought American technology into the common market.

Nevertheless, the individual decisions were being used to decide what should be said in the group exemption, and the final sentence may well have been inserted with a view to its construction.

Under the know-how regulation it seemed that only trivial know-how was excluded. These words have not been repeated in the new regulation, but it is thought that the Commission was only shortening the regulation and not changing the scope of its application.

The European Parliament wanted the definition in the know-how regulation to eschew the test of substantiality and include know-how 'capable of being exploited', i.e. that it should 'function' or that it should be a necessary or essential component in the manufacture of goods under licence.[10] In my view the test in both final regulations is broader and includes know-how on what does not work as well on what does. Often that is of great practical importance. It may be invaluable to avoid unfruitful avenues of r & d which may cost considerable sums and delay success. The test is expressed to apply *ex ante* when the agreement is made, and not *ex post* when its legality is in question. The definitions go far towards meeting the objections that it is not easy to say whether know-how is substantial or important.

In conclusion, I would regret the limitation of the regulation to know-how that is substantial, were the limitation not defined so narrowly. In so far as the permitted restrictions relate only to the licensed technology or to products made thereby,[11] they will be trivial if the technology is. Consumers will be able to find alternative supplies. The qualifications seem to me unnecessary. They lead to some uncertainty in enforcing contracts and may reduce the incentive to innovate and to grant or take a licence. Nevertheless, the definitions of secrecy and substantiality have been broad since the know-how regulation and should rarely cause problems.

If the know-how ceases to be secret and substantial and relevant patents have expired, the group exemption ceases to apply by virtue of article 1(2)–(4), discussed at 6.6.5 to 6.6.7 below, although where there was qualifying know-how, the provisions in Article 1(1)(7) and (8) may continue. Nevertheless, royalties may still be payable unless disclosure was through the action of the licensor or a third party[12]—it is only the restrictions of competition that may become void. Parties to a licence should consider whether to provide for the rest of the agreement to continue thereafter.

---

[10] See also explanatory statement, n. 4 above, 659.

[11] The only provisions exempted that may affect products not made by exploiting the licensed technology are the licensee's obligation to use his best endeavours to make and market the licensed product, exempted by art. 2(1)(17), and his obligation to pay a minimum royalty, etc., exempted by art. 2(1)(9). Both these provisions may deter him from using rival technology. The Commission, however, has power to withdraw the benefit of the exemption under art. 7(4) if the parties were already competing manufacturers when the licence was granted and the agreement included any of these provisions. See also 6.2 below.

[12] Recital 21.

6.1.1.3 'Identified' (Article 10(4))

To ensure that these conditions are fulfilled and that the know-how provides a valid justification for exempting territorial restraints and other provisions, the original know-how must be identified. Article 10 provides that:

> (4) 'identified' means that the know-how is described or recorded in such a manner as to make it possible to verify that it satisfies the criteria of secrecy and substantiality and to ensure that the licensee is not unduly restricted in his exploitation of his own technology, to be identified the know-how can either be set out in the licence agreement or in a separate document or recorded in any other appropriate form at the latest when the know-how is transferred or shortly thereafter, provided that the separate document or other record can be made available if the need arises;

It is likely to be more difficult to identify the know-how than to ascertain whether it is secret and substantial. It may be partly recorded in many different documents, and much of it may be in the head of the employee who uses and may have developed it. 'Show-how' that cannot be recorded was treated in earlier drafts of the know-how regulation as a service not a licence. It now seems to be sufficient that the know-how be recorded in writing or graphics and may include some 'show-how' too.[13] The record should be made either when the know-how is transferred or shortly afterwards.

There may be a practical problem keeping the records secret, which must increase the risks of unauthorized disclosure. Records may be locked up in a vault, provided that they can be produced later if a question arises about what know-how was included originally, or added subsequently. Confidentiality is inherent in know-how licensing. I doubt whether it suffices to record that the know-how includes a device for performing a particular function.

Where the licensee is required to obtain secret ingredients such as the bundle of herbs in *Campari*[14] or the pre-mix in *Rich Products/Jus-Rol* from

---

[13] In an individual decision, *Rich Products/ Jus-Rol*, cited n. 8 and described at 3.1.4 above, the Commission exempted a know-how licence when oral as well as written know-how was communicated, in part by reciprocal visits by the staff of the two parties (para. 11). Only the fact that the visits took place seems to have been recorded and not the details of what was done.

At the conference organized by Legal Studies and Services in Brussels in Apr. 1989, Signor Guttuso suggested that under a similar provision in the know-how reg., it might be safer to take a video record of the things shown and keep it in case the validity of terms in the licence is challenged in a national court or by the Commission.

[14] *Campari—Re the Agreement of David Campari Milano SpA* [1978] OJ L70/69, [1978] 2 CMLR 397, CMR 10,035.

the licensor, the know-how in those ingredients is not licensed and need not be identified. Nevertheless, for the regulation to apply, there must be relevant patents or other substantial and secret know-how that is identified and there may be none other than the secret formula.[15] To require pre-mixes to be recorded, even if locked away, would increase the risk of misappropriation.

The requirement of identification is limited to its purpose. The question arises whether it suffices to record enough know-how to qualify as secret and substantial, or whether it is necessary also to record anything that is substantial in order to rebut the licensee's claim that it developed it itself and is entitled to exploit it outside the licence.[16]

The Commission might require to see the records when deciding whether an agreement is exempt under the regulation. If the secrecy of the know-how is important, one might not send the record to the Commission, but make it available for its officials to inspect, perhaps in a lawyer's office in Brussels, and ensure that no photocopies are taken by the officials. The rights of the Commission under articles 11 to 14 of regulation 17 must be limited by the doctrine of proportionality.

Commission and national officials are bound by a general obligation to keep *le secret professionnel* secret. According to civil law concepts, this extends beyond trade secrets to information which officials receive solely by virtue of their office.[17]

The more recent practice of the Commission, under which confidential documents are kept in a separate part of the file and it is more difficult to get into DG IV without being escorted to one's destination, has been successful in radically reducing the number of cases where confidential information has been wrongfully disclosed, but there is no reason why officials should be entitled keep the details of secret recipes and other formulae. Often the know-how will be very bulky and this may help to protect its confidentiality.

Since sufficient identification of the know-how is a condition of exemption, national courts may have to determine whether the records kept are sufficiently detailed. They do this already where the recipient claims that it was already familiar with the know-how licensed to it.

The Member States seem to have required the Commission to

---

[15] Commission officials have suggested that in *Campari* not only was there insufficient know-how communicated to the licensees, the know-how was probably ancillary to the trade mark and not *vice versa*, so the reg. might not apply to such a licence even today. See, however, my comment on art. 5(1)(4) at 5.1.4 above.

[16] The European Parliament recommended that the know-how should be described as fully as possible. See the explanatory statement, n. 4 above, 655.

[17] It is excellently analysed by **C. S. Kerse**, *EC Antitrust Procedure* (3rd edn. London: Sweet & Maxwell, 1994), at 8.18. See also *AKZO Chemie BV and AKZO Chemie UK Ltd* v. *Commission* (53/85) [1986] ECR 1965, [1987] 1 CMLR 231, CMR 14,318.

compensate for know-how being inherently less precise than a patent—there is no patent specification or claims—by requiring that the know-how be recorded in sufficient detail.

6.2 HORIZONTAL AGREEMENTS ARE LARGELY EXCLUDED BY ARTICLES 5 AND 7(4)

The Council did not limit the Commission's power to grant group exemptions to cover only vertical situations.[18] Where the licensee could have produced the object of the technology licensed without help from the licensor, the agreement would be treated as horizontal in the United States,[19] and the group exemption should not apply, even if individual exemptions might be appropriate in the absence of market power or where there are efficiencies. This policy consideration has been met only in part by article 5(1), which excludes some horizontal agreements.[20] Moreover, article 7(4) enables the Commission to withdraw the exemption where the parties were already competitors before the grant of the licence and the licensee's obligation to produce a minimum quantity or to use his best endeavours have the effect of preventing the licensee from using competing technologies. The obligation to perform a minimum number of operations is not specifically listed, but this is not important in as much as the list in article 7 is not exhaustive. These are the only provisions permitted that are not limited to the licensed technology or products made thereby, and could be used to discourage a competitor from acquiring rival technology.

Bronbemaling v. *Heidemaatschappij*,[21] was discussed at 5.1.1.1.2 above. The Commission disapproved of licences granted to three firms that had opposed the grant of a patent on terms that no further licences would be granted, save to the holder's subsidiary, without the consent of the majority of licensees. At first sight, this may be perceived as an obligation of the same type but more limited in scope than the exclusivity permitted by article 1(1), and so exempted by article 1(5). Although the licence to each firm seems to have been bilateral, the applicant for the patent and the three opposers to whom the licences were granted were already competing and using the invention. It may well be that such an agreement would come within the exclusion of technology pools under article 5(1)(1).

Suppose, however, that there had been only one firm opposing the grant and it had been granted an exclusive licence. There would be no reciprocal

---

[18] In reg. 19/65 [1965] OJ Spec. Ed. 35.   [19] See 1.4.2 above.
[20] See 5.1.1–5.1.1.3 above.
[21] [1975] OJ L249/27, [1975] 2 CMLR D67, CMR 9776.

rights to bring the licence within the exclusion of article 5(1)(3). So it would be exempt unless a bilateral agreement can be described as a technology pool within the meaning of article 5(1)(1). The Commission would have to take a formal decision under article 7 to withdraw the exemption, especially if the licensee had a significant market share at the date the licence was granted.

6.3 ANCILLARY PROVISIONS RELATING TO INTELLECTUAL PROPERTY RIGHTS
(RECITAL 6 AND ARTICLES 1(1) AND 5(1(4)))

At 5.1.4 above, the exclusion from the regulation of ancillary provisions relating to trade marks and other intellectual property rights was considered and the unusual definition of 'ancillary provisions' in article 10(15) noted. Pure trade mark, copyright, and software licences have remained outside the regulation, although the technical information communicated may include some software or engineering drawings protected by copyright without preventing the application of the group exemption.

There may be some difficulty in deciding whether licences under these rights are 'of assistance in achieving the object of the licensed technology' within the meaning of recital 6. This formulation is, however, easier to apply than a concept of ancillary restraints, which surfaced under the patent regulation in *Boussois/Interpane*.[22] Nevertheless both recital 6 and the introductory words of article 1(1) extend the ambit of the regulation to include ancillary provisions relating to trade marks and other intellectual property rights.

If these words are treated as exclusionary rather than expansive, we may have to advise whether the trade mark, software, or other licence is 'of assistance in achieving the objective of the licensed technology' or *vice versa*. This will often be impossible: frequently both are essential.

I consider that the words are inclusive, and provided that the intellectual property rights assist the object of the licensed technology and that no restrictions are attached to the property rights that are not attached to the know-how and exempted by the regulation, it does not matter whether the intellectual property rights are of less value than the technology or not. I have also argued, in the alternative, that a provision may be ancillary if required to make the technology licence commercially viable. Either view would remove the need for sophistry. Until this is established by case law,

---

[22] Cited at n. 6 above and described at 3.1.3 above. The Commission did not analyse the words of the recital, but concluded in para. 20 only that the patent reg. did not apply because the know-how was crucial. It proceeded to grant an individual exemption.

the cautious draftsman of know-how licences may highlight the importance of the know-how and the assistance in selling the product provided by the trade mark.

6.4 TO WHICH ONLY TWO UNDERTAKINGS ARE PARTY

Council regulation 19/65, under which all three Commission regulations were made, requires the exemption to be limited to agreements to which only two undertakings are party. 'Undertaking' is a term broader than 'person'. It is much the same as the concept of 'firm' used by economists. It includes any collection of resources for commercial purposes, including services as well as goods. Where several legal persons form a single economic entity, the Community Court held in *Hydrotherm* v. *Andreoli*[23] that they form a single undertaking. Article 6(3), with the definition of 'connected undertakings' in article 10(14), also provides that the rights or obligations may be accepted by other companies within the group. See 5.2.3 above, where the Court's case law on relations between members of the same corporate group was considered.

---

[23] *Hydrotherm Gerätebau GmbH* v. *Compact del Dott. Ing. Mario Andreoli & C Sas* (170/83), [1984] ECR 2999, [1985] 3 CMLR 224, CMR 14,112. Signor Andreoli controlled a company and a partnership, and all three were party to an agreement with their exclusive distributor for the Federal Republic. It was held that this was an agreement between only two undertakings and capable of being exempted by reg. 67/67 ([1967] OJ Spec. Ed. 10) which was made under the same *vires* as the know-how reg.

The Court has not specified how dependent one person must be on the other for them to constitute a single undertaking. The actions of a half-owned subsidiary were attributed to *Commercial Solvents*, which appointed the chairman who had a second or casting vote. The policy in question was determined by Commercial Solvents: *Istituto Chemioterapico Italiano SpA and Commercial Solvents Corporation* v. *Commission* (6 & 7/73), [1974] ECR 223, [1974] 1 CMLR 309, CMR 8209. The acts of a subsidiary that were contrary to its parent's instructions were not attributed to the parent in *BMW Belgium SA and Others* v. *Commission* (32 & 36–82/78), [1979] ECR 2435, [1980] 1 CMLR 370, CMR 8548. The Court referred to the concept of dependence in a different context in *Centrafarm BV and Adriaan De Peijper* v. *Sterling Drug Inc.* (15/74) [1974] ECR 1174, [1974] 2 CMLR 480, CMR 8245:

'41. Article 85, however, is not concerned with agreements or concerted practices between undertakings belonging to the same concern and having the status of parent company and subsidiary, if the undertakings form an economic unit within which the subsidiary has no real freedom to determine its course of action on the market, and if the agreements or practices are concerned merely with the internal allocation of tasks as between the undertakings.'

In *Viho Europe BV* v. *Commission*, (T–102/92), [1995] ECR II–217, [1995] 4 CMLR 299, [1995] 1 CEC 562, appeal filed (73/95P), the Court of First Instance confirmed the Court's case law and extended it to export bans imposed by a parent on its subsidiary.

See also 5.2.3, where the concept of 'connected undertakings' is described.

6.5 ONE OR MORE OF THE FOLLOWING OBLIGATIONS

To come within the group exemption, the parties, or one of them, must accept at least one of the restrictions listed in article 1(1). The restriction need not be accepted to its full extent. Article 1(5) provides that the exemption shall also apply where the parties undertake obligations of the same type but with more limited scope. Most of the obligations relate to exclusive licences and associated territorial protection, but the final two items are different.

It is wrong to assume that all the agreements exempted by the regulation will be exclusive. It is possible to impose sales and manufacturing restrictions on a non-exclusive licensee in a commercially desirable territory in order to persuade a firm to take an exclusive licence in a territory that is less rewarding. Moreover, article 1(1)(7) exempts an obligation to use the licensor's trade mark and article 1(1)(8) exempts a licence to manufacture unlimited quantities for the licensee's own use.

6.6 TERRITORIAL PROTECTION THAT IS PERMITTED AND LICENCES LIMITED TERRITORIALLY

As explained at 2.2.1 to 2.2.5 above, in *Nungesser*[24] the Court ruled that open exclusivity does not, in itself, infringe article 85(1): an obligation on the licensor of plant breeders' rights not to grant further licences for or itself exploit the technology within the territory of the licensee may be justified.

As described at 1.4.1 above, the Court has gone further than the Commission in ruling that various ancillary restraints leading even to absolute territorial protection that are essential to enable the holder of copyright to calculate and obtain royalties for every performance of a film do not infringe article 85(1) in *Coditel II*.[25] Similarly in relation to plant breeders' rights in basic seed in *Erauw-Jacquéry* v. *La Hesbignonne*,[26] the

---

[24] *Nungesser (L. C.) AG and Kurt Eisele* v. *Commission* (258/78) [1982] ECR 2015, [1983] 1 CMLR 278, CMR 8805, discussed at length at 2.2 ff.
[25] *Coditel II—Coditel SA, Compagnie Général pour la Difussion de la Télévision and others* v. *Ciné-Vog Films SA and Others* (262/81), [1982] ECR 3381, [1983] 1 CMLR 49, CMR 8862; *Warner Bros Inc. and Metronome Video ApS* v. *Erik Vieuff Christiansen* (158/86), [1988] ECR 2605, [1990] 3 CMLR 684, [1990] 1 CEC 33, where the right of the holder in the country of import to control leasing was not exhausted by the sale in the UK.
[26] *Erauw-Jacquéry (Louis) SPRL* v. *Société La Hesbignonne SA* (27/87), [1988] ECR 1919, [1988] 4 CMLR 576, [1989] 2 CEC 637. See 1.4.1, and 2.2.5, text to nn. 48 and 49 and 2.2–2.2.5 above.

Court ruled that absolute territorial protection might be needed to enable the holder to control the propagation of certified seed.

At 3.1 to 3.1.8, I explained how narrowly the Commission interpreted these judgments in its later decisions, and at 3.1 I suggested that this may be explained by the Commission's desire to gain experience through some exempting decisions on which it might base its group exemption. The Commission has now interpreted 'open exclusivity' more widely in recital 10, which states:

> (10) Exclusive licensing agreements, i.e. agreements in which the licensor undertakes not to exploit the licensed technology in the licensed territory itself or to grant further licences there, may not be in themselves incompatible with Article 85(1) where they are concerned with the introduction and protection of a new technology in the licensed territory, by reason of the scale of the research which has been undertaken, of the increase in the level of competition, in particular inter-brand competition, and of the competitiveness of the undertakings concerned resulting from the dissemination of innovation within the Community. In so far as agreements of this kind fall, in other circumstances, within the scope of Article 85(1), it is appropriate to include them in Article 1 in order that they may also benefit from the exemption.

The Commission limits its view that open exclusivity does not, in itself, infringe article 85(1) to technology that produces a new product. In *Nungesser*, the Court referred to the newness of INRA's variety, but it interpreted 'newness' more widely than the Commission, which considered that the product was old, since maize has been known for generations. The Court observed that the INRA varieties were a great improvement on the old, in that they could be grown in the colder climate of Northern Europe.

In my view, the Court mentioned the novelty of the variety as a relevant factor, but only as one of a non-exhaustive list of reasons for investment being required that would justify open exclusivity. The variety was new. The exploitation of something not new but recently found to be useful, or for which a cheaper method of manufacture had been discovered, might also require investment. It is clear from the Court's judgment in *Nungesser* that an improvement on something already known may qualify as sufficiently novel.

Where *Nungesser* clearly applies, there may be no need to rely on the regulation, but in view of the Commission's narrow interpretation in its decisions adopted before preparing the know-how regulation, there can be few licences to which it clearly applies, so most advisers will consider increasing legal certainty by bringing their licences within the regulation when this can be done without serious commercial loss.

The first seven sub-paragraphs of article 1(1) exempt exclusive licences, ancillary territorial protection, and the use of trade marks in terms almost identical to the earlier licensing regulations. Article 1(1)(8) was first introduced in the know-how regulation. Article 3(3) and (7), which circumscribes the territorial protection closely, is also almost identical to article 3(12) and (11) of the know-how regulation.

At no point does the regulation state that the licensor may promise one licensee to limit the exploitation of other licensees in the ways permitted by article 1. Nevertheless, recital 12 asserts that desirable effects flow from the dissemination of technology made possible by exclusive licences and limited export bans. The reasoning is not explained very fully. Nevertheless, since the Commission states that territorial protection may be necessary to induce investments, it may be inferred that a promise to impose the permissible restrictions is also exempted. It may also be argued from the reference to 'fair competition' in the preamble to the Treaty that the promise to bind other licensees is the counterpart to the acceptance by each licensee of similar restraints.

6.6.1 Absolute Territorial Protection as Between Licensor and Licensee

Article 1(1)(1)–(3), as qualified as to duration by article 1(2)–(4), exempts restrictions on the grant of further licences in the licensee's territory, and on the licensor and licensee exploiting the licensed technology in the other's territory for ten years after any licensee first put the protected products on the market within the common market, or for as long as there is a relevant patent in both countries.[27] The first three sub-paragraphs exempt provisions that licensor and licensee may not exploit the technology in the territory of the other:

(1) an obligation on the licensor not to license other undertakings to exploit the licensed technology in the licensed territory;
(2) an obligation on the licensor not to exploit the licensed technology in the licensed territory himself;
(3) an obligation on the licensee not to exploit the licensed technology in the territory of the licensor within the common market.

Some of these words are defined in article 10. 'The licensed territory' is defined in article 10(11):

(11) 'the licensed territory' is the territory covering all or at least part of the common market where the licensee is entitled to exploit the licensed technology;

---

[27] The commencement of the permitted duration cannot be earlier and is usually later than under the earlier regulations. See 6.6.5–6.6.7 below.

Exploitation is defined disjunctively in article 10(10):

> (10) 'exploitation' refers to any use of the licensed technology in particular in the production, active or passive sales in a territory even if not coupled with manufacture in that territory, or leasing of the licensed products;'

The empowering regulation permits a territory only for a defined part of the common market, although in *Hydrotherm* v. *Andreoli*,[28] the Court ruled that a territory extending beyond the common market may benefit from the group exemption for exclusive distribution, made under the same empowering regulation. Recital 7 of the technology transfer regulation adds that it applies also to a territory wholly outside the common market. I still have a slight doubt about the validity of the regulation in so far as it permits a single territory for the whole common market. Usually, such a doubt can be avoided at little commercial cost by excluding some significant part of the common market, such as Luxembourg, while permitting the licensee to sell there on a non-exclusive basis.[29]

Many sole and exclusive licences may need no exemption where they are justified by the commitment of both parties to invest or by actual investment. Such licences may well qualify as open.[30] Nevertheless, it is seldom possible to give firm advice on this. So the regulation is helpful. It is also helpful when there are other territorial restraints imposed. Moreover, as stated at 2.2.3, the Commission has never considered the products sufficiently new to qualify as an open exclusive licence, so the regulation creates greater certainty.

I questioned at 6.2 whether a restriction on granting further licences without the consent of several licensees qualified as an obligation of the same type but more limited scope exempted by article 1(5). Such a promise is likely to have a horizontal element and may well be part of a patent or know-how pool excluded from the regulation by article 5(1)(1) or (3).[31]

An obligation on the licensee not to exploit the technology in territories reserved for the licensor may be imposed even if the licence is not exclusive.[32] The introductory words refer to an agreement under which at least one of the following obligations is accepted. Where some territories are commercially more desirable than others, it may be desired to protect the licensees in the less desirable territories from competition by non-exclusive licensees in the more desirable ones.

---

[28] N. 23 above, and recital 4, analysed at 4.2 above.

[29] The part excluded should be sufficient to permit parallel trade: see *Hydrotherm* v. *Andreoli*, n. 23 above.

[30] See *Nungesser*, n. 24 above, and the other cases described at 2.2–2.2.5 above.

[31] See *Bronbemaling* v. *Heidemaatschappij*, n. 21 above, discussed at 5.1.1.1.2 above.

[32] Art. 1(7)(12) of the know-how reg. defined these as territories 'expressly' reserved for the licensor. That definition has been abrogated in the technology transfer reg. The question arises whether it remains necessary to reserve the territories expressly in the licence.

6.6.2 Protection as between Licensees from Manufacture and Active or Passive Sales (Article 1(1)(4)–(6))

Article 1(1)(4) permits the licensor to impose restrictions on manufacture by each licensee in the territory of other licensees. Where the licence includes qualifying know-how, this may last for ten years from the date when the products were first put on the market within the EC by any licensee or as long as there is a necessary patent in both the licensed territory and that to.be protected.[33] Article 1(1)(5) permits a restriction of active sales for the same periods and article 1(1)(6) a restriction on passive sales for five years from the same date.

Article 1(1) continues to exempt:

(4) an obligation on the licensee not to manufacture or use the licensed product, or use the licensed process, in territories within the common market which are licensed to other licensees;

(5) an obligation on the licensee not to pursue an active policy of putting the licensed product on the market in the territories within the common market which are licensed to other licensees, and in particular not to engage in advertising specifically aimed at those territories or to establish any branch or maintain an [sic] distribution depot there;

(6) an obligation on the licensee not to put the licensed product on the market in the territories licensed to other licensees within the Common Market in response to unsolicited orders.'

These provisions have not changed from the earlier regulations. As previously, article 1(5) provides that licensor and licensee may accept the specified obligations protecting each other and other licensees, or obligations of similar effect but narrower scope. The focus is on the undertaking accepting the restriction, not on the person to be protected.

The policy reason for allowing a restriction on passive sales is not spelled out. Under regulation 1983/83, exclusive distributors may be restrained indefinitely from making active sales or from establishing a branch or depot outside their territories, but cannot be restrained from accepting unsolicited orders from outside. Under the technology transfer regulation active sales are defined slightly differently, and such a restraint is limited to the ten years or the life of the patent in both the territory protected and the territory licensed. On the other hand, a restriction on passive sales is permitted for five years from the time when any licensee first puts the product on the market within the common market. I assume that protection from passive sales is permitted because a licensee frequently has

---

[33] The permissible duration of these obligations is prescribed in art. 1(2)–(4) and considered at 6.6.5–6.6.7 below.

not only to develop the market within its territory, but also to invest in setting up a production line and, possibly, improving the technology.

The obligations that may be accepted are confined to action by the licensor and licensees: restrictions cannot be placed on their customers, directly or indirectly. Where freight is not an important element in the delivered price of a product, the restraints on exploiting the technology permitted may give very little protection since, even if there is a mixed licence including patent protection, a purchaser from one of the parties in the common market may sell wherever it likes in the common market, as a result of the doctrine of exhaustion developed by the Court in the early 1970s and described at 2.1.2 to 2.1.3 above.

Nevertheless, article 2(1)(14), discussed at 7.3.14 below, enables the licensor to reserve the right to exercise patent rights to prevent sales by one licensee directly into the territory of another. This seems to be a new interpretation of the doctrine of exhaustion narrower than that formerly accepted by the Commission. See 3.2 to 3.2.5 above. It may be useful when the products do not pass through dealers. It is not subject to any limitation of time other than the life of the patent in the country to be protected.

I regret the Commission's refusal to permit contractual restrictions downstream on the licensee's customers,[34] when the licence is vertical. The licensor has an incentive to grant no more territorial protection than it thinks is necessary to induce risky investment in tooling up and making a market. Since the Commission cannot analyse a market realistically in a group exemption, it might be better, even when there is no industrial property right to enforce, to rely on the licensor's interest in minimizing the margin enjoyed by each licensee than to limit the territorial protection that may be given, at least when the parties would not have been competitors but for the licence.

Where freight is an important cost, the licensee may be given far more effective protection than where it is not. The Commission seems to fear that this regulation may be used by competitors to divide the market.

6.6.3 Licensee's Obligation to Use the Licensor's Mark (Article 1(1)(7))

The introductory words of article 1 expressly include 'those agreements containing ancillary provisions relating to intellectual property rights other than patents'. Whether or not the licence is exclusive and whether or not

---

[34] To qualify, the licence must be bi-partite, but the licensor might have been allowed to require the licensee to impose a restriction on its customers. At 4.2.3 above I mentioned the difficulty currently arising when a licence is given to an East European licensee in restraining it from forming a forwarding post, independently managed, to sell the goods back into the common market.

there are territorial restrictions, the licensee may, by virtue of article 1(1)(7) be required to use the licensor's mark or get-up. When analysing article 5(1)(4) at 5.1.4 above, I questioned whether *Moosehead Whitbread*[35] would be decided the same way under the technology transfer regulation: whether, in view of the unusual definition of 'ancillary provision' in article 10(15), the provisions relating to intellectual property rights other than patents must be less valuable than the patents. If not, the regulation may possibly include production franchises even if the trade mark licence is crucial, provided there is also a patent or some substantial, secret, and recorded know-how licensed. I suggested that it might help to highlight the technological side of the licence and refer to the provisions about the use of the mark as ancillary.

At 5.1.4 above, I described the formalistic view of René Joliet that franchising is different from trade mark licences, but this does not seem to be accepted by officials.

From 5.1.2 above, I considered whether pure sales licences are entirely excluded from the ambit of the regulation by recital 5 and article 5(1)(5) or only from the territorial protection as by article 1(5) of the know-how regulation. If, contrary to the literal wording, the latter be correct, a pure sales licence with no territorial protection may qualify under the regulation if there is an obligation of the kind exempted by article 1(1)(7). It exempts:

> (7) an obligation on the licensee to use only the licensor's trademark or get up to distinguish the licensed product during the term of the agreement, provided that the licensee is not prevented from identifying himself as the manufacturer of the licensed products;

This exempted obligation may be without any limit of time, provided that the know-how remains secret and substantial or a necessary patent must remain valid in both the territory licensed and the territory protected.[36] The failure to exclude horizontal agreements from the scope of the group exemption has created difficulty. If the parties could have competed in a concentrated market without a licence, the obligation to use the licensor's mark and get-up may prevent the licensee from effectively developing its own identity. Where, however, the transaction is vertical, it is doubtful whether the requirement that the licensor's mark and get-up be used restricts competition, especially since the licensee must be permitted to indicate that it is the manufacturer.

Care must be taken in a trade mark licence to avoid the provisions in article 3(4) blacklisting the allocation of customers and article 3(5)[37] blacklisting maximum quantity restrictions. Article 3(4)[38] is expressly

---

[35] (90/186/EEC), [1990] OJ L100/32, [1991] 4 CMLR 391, [1990] 1 CEC 2127.
[36] See art. 1(2) and the last para. of art. 1(3).
[37] Analysed at 7.3.12.1 below.   [38] Analysed at 7.3.8.2 below.

limited to permit a requirement to mark the goods exempted by article 1(1)(7) and a field-of-use restriction is cleared by article 2(1)(8). Although article 1(1)(7) takes precedence over the blacklisted clause, it may be construed narrowly to permit obligations relating to the physical appearance of the product or its packaging, but there is some danger that article 3(4) or (5) may apply when the amount to be placed in a single package is specified. Such specification may serve to divide the wholesale and retail trades.

Unwanted quality specifications are subject to the opposition procedure. The registered-user agreement under which the mark will have to be licensed[39] will have to specify the goods or services to which the mark may be attached. It will almost certainly be treated as part of the same agreement as the technology licence.[40] Reciting that the quality specification is wanted by the licensee[41] may help to shift the onus of proof.[42]

The qualification that the licensee should be permitted to indicate that it is the manufacturer is explained by recital 10 of the patent licensing regulation, not repeated in the technology transfer regulation. The licensee will be less dependent on the licensor's mark when the licence terminates if it can state that it is the manufacturer during the currency of the licence.

6.6.4 Use Licenses (Article 1(1(8))

Article 1(1)(8) was added at a very late stage of the consultations on the know-how regulation. It is also without limit of time, subject to the know-how remaining substantial and secret or there being a patent both in the territory protected and that licensed. Article 1(1)(8) provides:

(8) an obligation on the licensee to limit his production of the licensed product to the quantities he requires in manufacturing his own products and to sell the licensed product only as an integral part of or a replacement part for his own products or otherwise in connection with the sale of his own products, provided that such quantities are freely determined by the licensee.

---

[39] At least under UK law, if the licensor wants to appropriate the benefit of the reputation earned by the licensee.

[40] See *BP Kemi/DDSF—Atka A/S* v. *BP Kemi A/S and A/S De Danske Spritfabrikker* [1979] OJ L286/32, [1979] 3 CMLR 684, CMR 10,165 at 3.1.5, n. 42 above.

[41] See art. 4(2) which submits unwanted quality specifications to the opposition procedure, analysed at 7.3.5.2 below, and note that in *Moosehead/Whitbread*, n. 35 above, the Commission seems to have treated the acknowledgement that a mark was valid, and a restriction on challenging it. See 7.3.6.2, text to n. 59 below.

[42] Note, however, that in *Moosehead/Whitbread*, n. 35 above, however, the commission treated the acknowledgement of the validity of the mark as a restraint on challenging its validity.

This was explained by recital 7 of the know-how regulation:

> (7) Both these [open exclusivity] and the other obligations listed in Article 1 encourage the transfer of technology and thus generally contribute to improving the production of goods and to promoting technical progress, by increasing the number of production facilities and the quality of goods produced in the common market and expanding the possibilities of further development of the licensed technology. This is true, in particular, of an obligation on the licensee to use the licensed product only in the manufacture of its own products, since it gives the licensor an incentive to disseminate the technology in various applications while reserving the separate sale of the licensed product to himself or other licensees.

The exemption of licences to use the technology for one's own needs marks a dramatic change from the attitude of the Commission and Court in *Windsurfing*.[43] Windsurfing International had applied for a patent in Germany covering both the rig and the board but, although proceedings were pending under German law, the Commission considered that the patent should be limited to the rig. The requirement that the licensees should sell the rigs they made only in conjunction with boards that they produced to a specification from Windsurfing could be perceived as a licence for use in making the complete sail vehicle. The Court did not refer to the term tie, but from paragraphs 54 to 59, it held that the obligation to sell the rigs only with approved boards restricted competition: it foreclosed other makers of boards. At paragraphs 38 to 53 the Court also condemned the obligation to use the rigs only with boards made to the specification in the licence agreement. The Court dismissed three justifications urged by Windsurfing International, but did not say why such a provision requires justification.

The Advocate General observed[44] that:

> (b) It is thus clear that the freedom of action of the licensees was restricted: they could not, without more, exploit the inventive thought in the context of its technical field of application, which concerns only one and not several separate markets. It is also clear, having regard to the scope of the German patent, that the restriction of their freedom of action with regard to the boards themselves was certainly not covered by the specific subject-matter of the patent and could therefore be caught by article 85(1) of the Treaty.

This formalistic theory,[45] which is implied also in the judgment at

---

[43] [1983] OJ L229/1, [1984] 1 CMLR 1, CMR 10,515, on appeal, *Windsurfing International Inc.* v. *Commission* (193/83), [1986] ECR 611, [1986] 3 CMLR 489, CMR 14,271.
[44] At 623 of the ECR.
[45] The idea that restrictions on the conduct of licensees which could not be restrained by use of the patent restrict competition is derived, in part, from the German cartel law and, in

paragraph 57,[46] is no longer fully accepted by the Commission which, following other judgments of the Court,[47] now believes that some restrictions outside the scope of patent protection may be necessary if innovation and its dissemination through licensing are to be encouraged.

In recital 24 of the technology transfer regulation,[48] the Commission stated that price and quantity restrictions were blacklisted because they seriously limit the extent to which the licensee can exploit the licensed technology and quantity restrictions may have the same effect as export restraints, but added that:

> This does not apply where a licence is granted for use of the technology in specific production facilities and where both a specific technology is communicated for the setting-up, operation and maintenance of these facilities, and the licensee is allowed to increase the capacity of the facilities or to set up further facilities for its own use on normal commercial terms. On the other hand, the licensee may lawfully be prevented from using the transferred technology to set up facilities for third parties, since the purpose of the agreement is not to permit the licensee to give other producers access to the licensor's technology while it remain secret or protected by patent (Article 2(1)(12)).

This is an example of the Commission returning towards the view stated in the 'Christmas Message'[49] that limited licences may not fall within the prohibition of article 85(1).[50] It seems, however, to have changed its mind in the final draft, and granted an exemption, although there are other examples of limited licences. At 3.2.4, we saw that 'exploitation' is defined disjunctively, so that the holder of technology may grant separate licences

---

part, from the doctrine of patent misuse in the USA. The latter is no longer accepted by the Department of Justice or the Federal Trade Commission which enforce the antitrust law, although the law may still be valid. See **James Venit**, 'In the Wake of *Windsurfing*: Patent Licensing in the Common Market', in **Barry Hawk** (ed.), *Thirteenth Annual Proceedings of the Fordham Corporate Law Institute* (New York: 1986) 517, at 522, and 1.4.1 above.

[46] The Court also concluded:
> '36. The clauses contained in the licensing agreements, in so far as they relate to parts of the sailboard not covered by the German patent or include the compete sailboard within their terms of reference, can therefore find no justification on the grounds of the protection of an industrial property right.'

[47] See 1.4.1 for the Court's development of the concept of ancillary restraints needed to make a transaction viable not being contrary to art. 85(1); 2.2–2.2.5 above for the Court's judgments on open and closed exclusivity; and 3.1–3.1.5 above for the Commission's views of restrictions outside the scope of the patent.

[48] Following recital 24 of an early draft of the know-how reg. [1987] OJ C214/2:
> quantity restrictions resulting from a license to cover the licensee's own requirements, as freely determined by the licensee, are no more than a use license which is not itself caught by Article 85(1).

[49] [1962] JO 2921, since withdrawn. See 3.2–3.1.2 above.

[50] The reg. does not go all the way, in as much as the quantities the licensee may use must not be limited. Indeed, limits of quantity are blacklisted by art. 3(5).

for manufacture, sale, or leasing, and possibly for licensing, but not for different geographic territories.[51] Other examples occur in the white list, which clears a ban on the use of the know-how after termination of the agreement[52] and some field of use restrictions,[53] although other customer restrictions are blacklisted by article 3(4).

6.6.5 The Duration of Territorial Protection and Other Exempted Provisions in Pure Patent Licences

Article 1(2) limits the duration of the exemption of pure patent licences;

> 2. Where the agreement is a pure patent licensing agreement, the exemption of the obligations referred to in paragraph 1 is granted only to the extent that and for as long as the licensed product is protected by parallel patents, in the territories respectively of the licensee (points (1), (2), (7) and (8)), the licensor (point (3)) and other licensees (points (4) and (5)). The exemption of the obligation referred to in point (6) of paragraph 1 is granted for a period not exceeding five years from the date when the licensed product is first put on the market within the common market by one of the licensees, to the extent that and for as long as, in these territories, this product is protected by parallel patents.

'Parallel patents' are defined in article 10(13) as:

> (13) . . . patents which, in spite of the divergences which remain in the absence of any unification of national rules concerning industrial property, protect the same invention in various Member States;

The claims and the legal definition of the rights may differ somewhat in the different Member States. Nevertheless, under the influence of the European Patent Convention, all the Member States now grant patents for twenty years from application and the other rules are becoming harmonized without the need for any harmonization directives. Moreover, European patents must be recognized and enforced in all Member States of the EC even when the criteria for grant are not identical.

The exemption lasts only if there remains a patent both in the territory licensed, or the agreement would not be a patent licence, and in the territory protected, as was the case under the former regulations. Sometimes a territory extends beyond a single Member State and there may be a patent in only one of these. The regulation does not specifically

---

[51] Since the doctrine of exhaustion, described at 2.1.2 and 2.1.3 above, prevents the use of intellectual property rights in one Member State to restrain imports from another where the product was sold by or with the consent of the holder, the Commission treated a territorial limitation as being outside the scope of the patent.

[52] Art. 2(1)(3). See other examples at 3.2.5, n. 64 above.

[53] Art. 2(1)(8).

address the problem of the availability or duration of the exemption where there is a valid patent only in parts of the licensed territory or that to be protected. Article 1(2) says 'for as long as the licensed product is protected' in both territories. I would argue that it is so protected if there is a patent in part of each territory.

The period of protection for an obligation not to make passive sales is different, and the same for all kinds of licence, pure patent, pure know-how, or mixed. It is five years from the date the products were first put on the market within the common market by any licensee. Where there is no qualifyig know-how, it will also cease when there remains no patent protection in either the licensed territory or that to be protected. Marketing by the licensor no longer counts as it did under the patent regulation. Nor does the date of the first licence within the EC as under the know-how regulation. The period now permitted will, therefore, usually be longer than under the earlier regulations and cannot be shorter. Nevertheless, the Commission has not followed the view of the Economic and Social Committee that the protection from passive sales should be as long as for protection from active sales.[54]

Where the licence is confined to a territory outside the common market, and the licensor supplies the whole common market itself without licensing anyone to produce and sell there, the period permitted for restraints on passive sales may never begin to run.

Note that, unlike licences that include qualifying know-how, the time limit applies to the provisions exempted by Article 1(1)(7) and (8) as well as to the territorial protection.

6.6.6 Duration of Territorial Protection in Pure Know-how Licences

Recital 13 of the technology transfer regulation explains the reason for limiting the obligations accepted in pure know-how licences:

> (13) Since the point at which the know-how ceases to be secret can be difficult to determine, it is appropriate, in respect of territories where the licensed technology comprises know-how only, to limit such obligations to a fixed number of years. Moreover, in order to provide sufficient periods of protection, it is appropriate to take as the starting-point for such periods the date on which the product is first put on the market in the Community by a licensee.

---

[54] Additional Opinion on the draft Commission reg. on the application of art. 85(3) of the Treaty to certain categories of know-how licensing agreements, Economic and Social Committee, adopted at the 254th plenary session on 23 Mar. 1988 [1988] OJ C134/10, [1988] 4 CMLR 498, 2.4.8 thereof.

The maximum period of territorial protection permitted for a pure know-how licence is prescribed by article 1(3):

> 3. Where the agreement is a pure know-how licensing agreement, the period for which the exemption of the obligations referred to in points (1) to (5) of paragraph 1 is granted may not exceed ten years from the date when the licensed product is first put on the market within the common market by one of the licensees.·

The 'licensed products' are defined in article 10(8) as:

> goods or services the production or provision of which requires the use of the licensed technology;

'Licensed technology' is defined in article 10(7):

> (7) 'licensed technology' means the initial manufacturing know-how or the necessary product and process patents, or both, existing at the time the first licensing agreement is concluded, and improvements subsequently made to the know-how or patents, irrespective of whether and to what extent they are exploited by the parties or by other licensees;

Consequently the duration of the territorial protection cannot be prolonged by adding improvements to the licence. Recital 14 explains:

> (14) Exemption under Article 85(3) of longer periods of territorial protection for know-how agreements, in particular in order to protect expensive and risky investment or where the parties were not competitors at the date of the grant of the licence, can be granted only by individual decision. On the other hand, parties are free to extend the term of their agreements in order to exploit any subsequent improvement and to provide for the payment of additional royalties. However, in such cases, further periods of territorial protection may be allowed only starting from the date of licensing of the secret improvements in the Community, and by individual decision. Where the research for improvements results in innovations which are distinct from the licensed technology the parties may conclude a new agreement benefitting from the exemption under this Regulation.

It is not easy to distinguish improvements from innovations that are distinct from the licensed technology. Again, if further territorial protection be required, and the improvement is somewhat different, it may be worth while for both parties to acknowledge in the subsequent licence that it relates to a distinct innovation. The distinction will have to be drawn by the Commission when asked for a negative clearance or exemption for further territorial protection, or by a national court asked to enforce contractual terms. This is unfortunate, as parties need certainty before they commit resources to exploiting the improvements, not a chance to litigate.

Problems may arise when the technology is constantly being developed.

Whether or not the initial licensee obtains improvements, the period begins to run from the first sale by any licensee in the common market of products made by the original technology as updated. Recitals 14 and 21 makes it clear that, although the licence may continue to attract royalties after the initial period of ten years has expired, no further territorial protection will be permitted for improvements, even if new licences are granted to include them unless the new technology is distinct, save through an individual exemption.

The 'licensed technology' probably does not include secret intermediate products supplied by the licensor, such as the pre-mix in *Rich Products/Jus-Rol* or the herbs in *Campari*. Such know-how is not licensed, but retained by the licensor, who supplies the intermediate product.[55]

Where there is no licensee within the common market, the period of territorial protection within it from sales by licensees outside will never start. The territorial protection may last as long as the know-how remains secret and substantial.

Even where there is no patent, considerable investments may have to be made by a licensee to develop the technology and create demand. It may need to be protected from the licensor and other licensees taking a free ride. The investment may not be forthcoming without longer territorial protection than is permitted by the regulation. Recital 14 makes it clear, however, that this would be relevant only to the grant of an individual exemption for longer periods. The parties who notify are more likely to receive a comfort letter, which might not make the protection easily enforceable in a national court.

The final paragraph of article 1(3) provides that:

> The obligations referred to in points (7) and (8) of paragraph 1 are exempted during the lifetime of the agreement for as long as the know-how remains secret and substantial.[56]

This period may well outlast that permitted for territorial restraints, provided qualifying know-how subsists.

6.6.7 Duration of Territorial Protection in Mixed Patent and Licensing Agreements

Recital 16 provides that:

> (16) The exemption protection should apply for the whole duration of the periods thus permitted, as long as the patents remain in force or the

---

[55] [1988] OJ L69/21, [1988] 4 CMLR 527, CMR 10,956, and *Re the Agreement of David Campari Milano SpA* [1978] OJ L70/69, [1978] 2 CMLR 397, CMR 10035, discussed respectively at 3.1.4 and 3.1.2, above.

[56] These may be imposed in a pure patent licence provided that a patent remains valid in the territory licensed.

> know-how remains secret and substantial. The parties to a mixed patent
> and know-how licensing agreement must be able to take advantage in a
> particular territory of the period of protection conferred by a patent or
> by the know-how, whichever is the longer.

This limits only the territorial protection and, provided that qualifying
know-how subsists, not the provisions in article 1(1)(7) and (8).

Article 1(4) provides that

> 4. Where the agreement is a mixed patent and know-how licensing
> agreement, the exemption of the obligations referred to in points (1) to (5) of
> paragraph 1 shall apply in Member States in which the licensed technology is
> protected by necessary patents for as long as the licensed product is protected
> in those Member States by such patents if the duration of such protection
> exceeds the periods specified in paragraph 3.
>
> The duration of the exemption provided in point (6) of paragraph 1 may
> not exceed the five-year period provided for in paragraphs 2 and 3.
>
> However, such agreements qualify for the exemption referred to in
> paragraph 1 only for as long as the patents remain in force or to the extent
> that the know-how is identified and for as long as it remains secret and
> substantial, whichever period is the longer.

Where there is a necessary patent in both the licensed territory and that
protected by a restriction of exploitation outside a territory, the exemption
may last beyond the ten years permitted for a licence of know-how. This
should not be permitted when an irrelevant patent is added to the licence
solely to obtain longer protection. According to article 10(5) 'necessary
patents' are:

> (5) . . . patents where a licence under the patent is necessary for the putting
> into effect of the licensed technology in so far as, in the absence of such a
> licence, the realization of the licensed technology would not be possible
> or would be possible only to a lesser extent or in more difficult or costly
> conditions. Such patents must therefore be of technical, legal or
> economic interest to the licensee;

The definition prevents the parties from licensing an irrelevant, but more
recent, patent merely to prolong the territorial restraints. It is unlikely to
exclude patents included for good commercial reasons. Nevertheless,
where the licensee wants separate technology for different purposes the
territorial restraints imposed for one purpose cannot be prolonged by
reference to a patent required for the other.

Territorial protection lasting longer than permitted by article 1(2) to (4)
is blacklisted. See 7.3.20.2 below.

Like the European Parliament commenting on a draft of the know-how
regulation,[57] I question the need for these time limits, with the uncertainty

---

[57] See also explanatory statement, n. 4 above, 660.

they entail. They reduce the total income of the licensor, by increasing the risk to its licensee. This may discourage investment in developing technology and taking a licence.

Unless the licensee could, without the licence, have achieved as good a product as it has achieved with it, the agreement is vertical.[58] This is now accepted by officials in the Commission, although some fear that they may not be able to identify potential competitors. Where the licence is vertical, the licensor has no incentive to protect his licensees from each other or from itself unless it thinks that this is necessary to persuade them to invest in tooling up and making a market.

It is more likely to be right than the Commission, especially in a group exemption, since it operates in the particular market and is backing its judgement with the licence of the know-how. Moreover, the incentives to investment must be evaluated *ex ante* as at the date of the transaction, not *ex post* in the light of subsequent events.[59]

Where the licensee was not even a potential competitor before the licence, it may be argued that no competition which would have been possible is restricted and article 85(1) is not infringed by further territorial protection.

The time limits work particularly badly when there is a continuous stream of improvements.

6.6.8 Subsequent Licensees can be Given only Shorter Territorial Protection

The protection permitted for each territory dates from the same time, so a licensee whose licence is granted later than others will enjoy territorial protection for a shorter time. The subsequent licensee may, however, have similar costs in erecting a plant and developing a market in his territory,[60] in which event it may need the protection to start, as the Economic and Social Committee suggested when commenting on the draft of the know-how regulation,[61] from the date of its own licence, so that there will still be sufficient time left to take advantage of it when the licensee starts to sell.

The Commission stated, in relation to the patent regulation, that the difficulty of persuading licensees late in the chain to invest in tooling up and making a market would encourage holders of technology to license throughout the common market rapidly.[62]

---

[58] See 1.4.2 above. [59] 1.3.1 above.

[60] It may enjoy some advantages over the initial licensee, such as the possibility of sending industrial customers to see the products being made elsewhere.

[61] Additional Opinion, n. 54 above, at 2.4.2–2.4.4 thereof.

[62] See the Commission's press release in relation to the patent reg.: IP(84) 270.

This has two disadvantages. A large firm with manufacturing capacity can develop the technology in-house, and still have the possibility of giving protection to each licensee from active sales by the others for ten years after the first licence in the EC. A smaller innovator may not be able fully to develop the technology in-house and may need a licensee to get the bugs out of the system by producing on an industrial scale. The initial licensee may place the goods on the market before further licences are granted. The period for maximum protection available for subsequent licensees will start from that date. This may be thought unfair and may encourage the innovator to merge with the initial licensee.

A smaller innovator may consider granting such an initial development licence for a territory outside the common market, so as still to have the chance of giving ten years' protection against active sales and five against passive to the subsequent licensees which will benefit from the development carried out by the first licensee, but will still have to invest in tooling up and making a market. The advantage of receiving the fully developed technology may be counterbalanced by a duty to pay higher royalties than the first licensee or lump sums. This second scenario has the disadvantage of encouraging development to be done outside the common market. As the common market expands this becomes more and more inconvenient.

If the licensor prefers to grant a longer period of protection, an individual exemption must be obtained.[63] Otherwise, the licensee will have to try to earn higher profits in the initial periods while it can still be protected. The investment and risk cannot be amortized over the longer period. To make it worth while for the later licensees to invest, the licensor may have to charge reduced royalties. It may have to compensate for these by charging higher royalties to the earlier licensees.

The shortness of the periods of protection exempted increases the risk borne by licensees, especially those licensed at a late stage or for a territory that it is more expensive to exploit, and reduces the profitability of the innovation for the licensor. I would have preferred horizontal agreements to have been more fully excluded from the regulation and the periods of territorial protection to have been unlimited.[64] The innovator has no incentive to share any market power it may have with his licensees and if it restrains them from competing with each other it must be because of the need to induce investments by them.

---

[63] The Commission grants few formal exemptions because it has developed such an elaborate procedure therefor. The parties are more likely to receive a comfort letter stating that the agreement merits exemption. This may make it harder to enforce any additional territorial protection in national courts. See 1.1.1 above.

[64] See the views of the European Parliament, explanatory statement, n. 4 above, 660.

6.6.9 Licence Networks where there is Patent Protection in Some Member States Only

The same technology is unlikely to be protected by patent in some Member States and remain secret in others once the patent office publishes the specifications and claims. Know-how may, however, be complemented by 'necessary patents' in some Member States only. Where H has a patent in only some Member States, say Germany and the United Kingdom, the licences to G in Germany and U in the United Kingdom will be mixed patent and know-how licences, but if there are no patents in Finland and Portugal, licences to F in Finland and P in Portugal will be pure know-how licences. Assuming that the licences are granted when the patent still has much more than ten years to run, the duration of the exemption seems anomalous.

F and P can be required not to exploit outside their territories only for ten years and not to accept unsolicited orders for five years after the first sale in the common market by a licensee, after which they must be allowed to manufacture or sell throughout the common market, even in the United Kingdom and Germany. The licensees with patent protection may, however, be kept out of each other's territories for the duration of the patents in both countries, whichever is shorter, save for passive sales, for which the five-year cap applies to all the licensees but not to any territory reserved for the licensor. They can be kept out of Portugal and Finland only for the ten-year period, five for passive sales.

The regulation is focused on the licensee who may accept a restriction rather than on the licensee requiring protection. The position is as anomalous as that considered at 5.1.2.5 where the holder manufactures and uses distributors in some countries and grants technology licences in others.

6.6.10 Individual Exemption where Protection is Needed for Longer

Recital 14 expressly states that individual exemptions may be granted for a longer period when justified by expensive and risky investment where the parties were not competitors at the date of the grant of the licence.

6.7 TERRITORIAL PROTECTION IS CIRCUMSCRIBED (ARTICLE 3(3) AND (7))

Article 3(3) and (7) circumscribes the territorial protection permitted by article 1(1) and will be considered in Chapter 7 below.

6.7.1 Withdrawal of Exemption where Territories are Unduly Protected (Article 7(2) and (3))

Article 7 provides for the Commission to withdraw the exemption in certain circumstances.[65] Paragraph 3 instances conduct by one or both parties to hinder parallel exports, whether or not required or carried out as a concerted practice. See also 9.1 and 9.1.3.3 below.

6.8 DURATION OF LICENCE

The periods of territorial protection in all technology licences and of the provisions exempted by article 1(1)(7) and (8) in licences including qualifying know-how are limited as described at 6.6.5 to 6.6.8 above, and an agreement cannot automatically be prolonged by the unilateral addition of further technology. Nevertheless, there is no reason why the licence should originally be limited to ten years or to the period of patent protection. This was made clear by recitals 15 and 20 of the know-how licensing regulation. Recital 15 provided that the licensee should no longer have to pay royalties if the know-how becomes publicly known through the action of the licensor, other than the mere sale of a product that can be reverse-engineered. Royalties may, however, continue to be payable after the know-how has entered into the public domain through the action of third parties,[66] although, in that event, the exemption may eventually be withdrawn by the Commission under the general words of article 7. The words in article 7(7) of the know-how regulation to this effect have been removed, but probably only for the sake of simplicity.

Consequently, a licensor may pass some of the risk of premature disclosure on to the licensee. Often the licensee is a larger firm than the licensor and better able to bear the risk, in which event it may agree to continue paying royalties in return for other elements of the agreement as a whole. This has been confirmed by the Court's judgment in *Ottung* v. *Klee*.[67]

---

[65] In *Tetra Pak I* [1988] OJ L272/27, [1990] CMLR 47, CMR 11,015, discussed at 9.1.1 below, the Commission stated that such a withdrawal operates only from the date of the decision. Since the decision affects the rights of the parties, it is a formal act. It must be adopted by the Commission acting collegiately, the parties must be given a chance to submit their views to the Commission, and may appeal against the decision to the Community Court. The Commission has withdrawn a block exemption by individual decision only once, in *Langnese-Iglo* (93/406/EC), [1993] OJ L183/19, [1994] 4 CMLR 51, [1993] 2 CEC 2123. See 9.1.1. below.

[66] See recital 21 and art. 2(1)(7) and art. 3.

[67] *Ottung* v. *Klee & Weilbach A/S and Thomas Schmidt A/S* (320/87), [1989] ECR 1177, [1990] 4 CMLR 915, [1990] 2 CEC 674, where the Court refused to condemn an obligation to continue paying royalties after a patent had expired, in as far as they were in return for benefits obtained before expiry.

6.9 OBLIGATIONS OF THE SAME TYPE BUT MORE LIMITED SCOPE
(ARTICLE 1(5))

Article 1(5), like the patent regulation, permits obligations of the same type but more limited scope:

> (5) The exemption provided for in paragraph 1 shall also apply where in a particular agreement the parties undertake obligations of the types referred to in that paragraph but with a more limited scope than is permitted by that paragraph.

This prevents the application to the technology transfer regulation of the Commission's decision in *Junghans*[68] where it found that an agreement to supply only three distributors in one area did not fall with in the exemption of regulation 67/67, although the Commission proceeded to exempt the agreement by analogy to the regulation. It stated that regulation 67/67, which does not contain any provisions for exempting clauses of more limited scope, does not apply unless the exclusive protection is complete. The provision about obligations of more limited scope was becoming standard, but has been omitted from the group exemption for franchising. Its inclusion in this regulation is welcome. See also 9.1, 9.1.1, and 9.1.3.3 below.

It has been suggested, on the one hand, that clauses of more limited scope should apply *a fortiori*, since they are likely to have a less restrictive effect on competition. On the other hand, it may be argued that they have less effect in bringing about the improvements to distribution or production or technical or economic progress, envisaged by article 85(3). These arguments should balance, so I am delighted that such a provision has been included.

---

[68] *Junghans GmbH (Re the Agreement of Gebrüder)* (77/100/EEC), [1977] OJ L30/10, [1977] 1 CMLR D82, CMR 9912. See also *Spices: Brooke Bond Liebig* [1978] OJ L53/20, [1978] 2 CMLR 116, CMR 10,017, where the reseller agreed to give preference to the supplier's spices in defined ways, but not to purchase only from it.

7

The White and Black Lists
(Articles 2 and 3)

7.1 INTRODUCTION TO THE WHITE LIST (RECITAL 18 AND ARTICLE 2)

The general scheme of article 2 is the same as that of the white list in the two earlier regulations. It starts:

> 1. Article 1 shall apply notwithstanding the presence in particular of any of the following clauses, which are generally not[1] restrictive of competition.

Recital 18 explains that the provisions listed are commonly found in agreements and rarely restrict competition. They are exempted by article 2(2) to give greater legal certainty just in case they may be:

> (18) It is desirable to list in this Regulation a number of obligations that are commonly found in licensing agreements but are normally not restrictive of competition, and to provide that in the event that because of the particular economic or legal circumstances they should fall within Article 85(1), they too will be covered by the exemption. This list, in Article 2, is not exhaustive.

Article 2(2) gives legal certainty to qualifying licences by providing that:

> 2. In the event that, because of particular circumstances, the clauses referred to in paragraph 1 fall within the scope of Article 85(1), they shall also be exempted even if they are not accompanied by any of the obligations exempted by Article 1.

It is clear that the exemption of article 2(2) applies to pure and mixed patent and know-how agreements, whether or not one or more of the obligations listed in article 1 be accepted. Unlike regulations 1983/83 and 1984/83, granting a group exemption to exclusive distribution agreements, and the franchising regulation, no 4087/88, article 2(3) of the technology transfer regulation extends the exemption to obligations of the same type but more limited scope. This is welcome.

The Commission has not expressly been given power to grant group negative clearances, but the technique of providing a white list and actually exempting such provisions if they do infringe article 85(1), which is now

---

[1] Unfortunately, the Italian text omits the word 'not'. This is unlikely to cause trouble in the light of the other texts and of recital 18.

adopted in some group exemptions, may circumvent its lack of power. I have heard no one question its validity.[2]

Recital 18 and the introductory words of article 2(1) may be very helpful when justifying an agreement that does not come within the regulation, such as a pure copyright or design licence excluded by article 5(1)(4), because they assert that these terms rarely infringe article 85(1). Recital 18 states also that the list is not exhaustive. So, one may argue that different obligations with similar objectives escape the prohibition of article 85(1). This is confirmed, in that the list in article 2(1) is prefaced by the words 'in particular'. A payment in advance has much the same effect in not deterring the licensee from producing more, as does a minimum royalty.[3] Only the latter is mentioned in article 2(1)(9), but the payment in advance may also qualify.

Moreover, provided that there are no blacklisted provisions, a pure or mixed patent or know-how licence within the meaning of the introductory words of article 1(1) is actually exempted even if a whitelisted provision does happen to infringe article 85(1).

The order of the clauses in the know-how and technology transfer regulations is different from that in the patent one, and there are significant additional items this time. Some of the provisions are closely circumscribed by article 3, which lists the circumstances that prevent the regulation from applying to an agreement or by article 4(2), which provides the opposition procedure. Sometimes, only a stronger restriction is blacklisted. Nevertheless, the new regulation is simpler and easier to apply than its predecessors.

If an agreement includes a provision of intermediate strength, the question may arise whether it is cleared or exempted by the white list, which is expressed not to be exhaustive, or whether it is an unlisted provision that might be notified under the opposition procedure.[4] Articles 2, 3, and the specific examples in article 4(2) are most conveniently considered together. I shall start with the white list, considering the items in articles 3 and 4 with those in the white list which they limit, then address the remaining items of the black list.

---

[2] Compare recital 2 of the group exemption for certain categories of joint research and development agreements, reg. 418/85 [1985] OJ L53/5. It states that collaboration which does not extend beyond joint r & d to the exclusion of joint production seldom infringes art. 85(1).

[3] Payments in advance were cleared in *Boussois/Interpane* [1987] OJ L50/30, [1988] 4 CMLR 124, CMR 10,859, para. 22(c), although no reasons for doing so were given.

[4] According to which one may notify an agreement that qualifies as technology transfer, and which contains no blacklisted provisions, but at least one restriction of competition that is not listed. If the Commission does not oppose the exemption within 4 months, the agreement is exempted under the reg. The procedure is analysed in Ch. 8 below.

Article 3 provides that articles 1 and 2(2) shall not apply where one or more provisions listed therein is included in an agreement. Recital 19 states that:

> (19) The restrictions listed in Article 3 may fall under the prohibition of Article 85(1), but in their case there can be no general presumption that, although they relate to the transfer of technology, they will lead to the positive effects required by Article 85(3), as would be necessary for the granting of a block exemption. Such restrictions can be declared exempt only by an individual decision, taking account of the market position of the undertakings concerned and the degree of concentration on the relevant market.

According to the original published draft of the technology transfer regulation,[5] which did not include an opposition procedure, even when there was a blacklisted clause the regulation would apply to exempt other provisions in an agreement. It was only the blacklisted provisions that would be illegal and void if the agreement as a whole was contrary to article 85(1). This idea was dropped, however, when the opposition procedure was reintroduced. If there is a blacklisted clause, even if it does not restrict competition, the provisions listed in article 1(1) will not apply, although an individual exemption would theoretically be possible.[6]

After the adoption of the patent regulation, there was a time when, even in individual decisions, the Commission refused to exempt agreements that included a blacklisted clause, even when the reasons for blacklisting the provision did not apply, or did not apply in full. In *Velcro/Aplix*,[7] the Commission objected in an individual decision to a strong grant-back clause, at least when the basic patents had expired, although there was provision in the agreement for a reasonable royalty to be paid to the licensee and this was enforceable through arbitration.

> 57. The obligation to allow Velcro to acquire the title to patents in the Federal Republic of Germany, in the United Kingdom and in the

---

[5] [1994] OJ C178/3.

[6] More likely, the Commission would close its file, sending a comfort letter stating that the agreement was not forbidden by art. 85(1) or merited exemption. The procedure for granting exemptions is so cumbersome that it is rarely used.

[7] [1985] OJ L233/22, [1989] 4 CMLR 157, CMR 10,719; **Valentine Korah**, 'Comment, *Velcro/Aplix*' (1985) 7 EIPR 296. Compare **James Venit's** basic and perceptive comment; in 'In the Wake of *Windsurfing*: Patent Licensing in the Common Market', in **Hawk** (ed.), *Thirteenth Annual Proceedings of the Fordham Corporate Law Institute* (New York: 1986), 517, 534.

Netherlands for improvements discovered by Aplix is, in principle, an unwarranted extension of the licensed patents in that the licensor is using his industrial property rights to appropriate certain foreign patents covering improvement inventions that are wholly or partly the work of his licensee.

No further reason was given. The reason given in recital 24 of the patent regulation for blacklisting strong grant-back clauses is that they put the licensee at a competitive disadvantage. This did not apply since the licensee was to be paid the value of his innovation. The reason given in earlier individual decisions is that such a grant-back clause removes the incentive for the licensee to invest in innovation, but again that reason did not apply or did not apply fully. Fair remuneration fixed by an arbitrator may be less than could be negotiated in the absence of the clause. It may also be considerably less than could be obtained by the licensee exploiting its improvements itself. Nevertheless, the decision did not go into these points, but merely asserted, without giving reasons, that once the basic patents had expired a strong grant-back clause was an abuse of patent rights.

In an individual decision on rose breeders' rights, *Royon* v. *Meilland*,[8] the Commission also condemned a strong grant-back clause without giving reasons, even when the basic rights had not expired.

Bureaucracies have a natural tendency to develop pigeonholes and not to think whether the causes for concern exist in a particular case. Officials can then operate faster. Consequently, the view that it was not appropriate to exempt blacklisted clauses should not have caused surprise. Fortunately, many Commission officials are now rejecting such a rigid approach.

In *Boussois/Interpane*,[9] the parties stated that the licence to produce in the factory provided by Interpane for Boussois implied a restriction on exploiting the technology elsewhere. In the absence of patents in some Member States, this territorial protection was blacklisted by the patent regulation, yet the Commission did exempt the provisions. More recently

---

[8] Sometimes known as *Plant Breeders' Rights* (*Roses*) [1985] OJ L369/9, [1988] 4 CMLR 193, CMR 10,757. The Commission reasoned that a feed-back clause restricted competition because it restrained Royon from competing in the supply of mutations where he would otherwise be justified in so doing (para. 23(a)). Para. 27 gives two reasons: first, the clause reduced the incentive to improve production and, secondly, the grant-back clause is not justified by the intellectual property right because it does not concern the subject matter of the licensor's plant breeders' right, only of some other plant breeders' right that might be obtained for a good new mutation.

Seen *ex post*, after Royon had discovered the mutation, the Commission is clearly right. Meilland does not seem to have exploited the new variety very strongly. Nevertheless, in the light of this decision, some holders of such rights might employ someone to look after their roses, rather than acquire his services, so as to avoid an agreement between undertakings. That would certainly not be more competitive. (See 1.3.1 and *Viho Europe BV* v. *Commission* (T–102/92), [1995] ECR II–217, [1995] 4 CMLR 299, [1995] 1 CEC 562, appeal filed (73/95P) at 5.2.3 above.)

[9] Cited at n. 3 and described at 3.1.3 above.

still, in *Delta Chemie/DDD*,[10] it exempted territorial protection in a know-how licence for twenty years, longer than was permitted by article 1 of the draft for the know-how regulation available at the time and, consequently, blacklisted.

In the field of distribution, too, the Commission is no longer condemn-ing blacklisted clauses without thought. In *Fluke/Philips*,[11] the Commis-sion sent a comfort letter favourable to a reciprocal exclusive distribution agreement,[12] although these were blacklisted in regulation 1983/83. Where the causes for anti-trust concern about a blacklisted clause are met in an individual case, as by the arrangement for paying a royalty to Aplix for improvements, the clause may not be condemned, although it prevents the application of the group exemption even to other provisions.

7.3 THE LISTED PROVISIONS

In the technology transfer regulation the recitals have been shortened,[13] and we are not given reasons for listing as black or white many of the clauses with which we are familiar from the earlier regulation. Recitals 20 to 24, however, do deal with some aspects of the lists. Reference will be made to the recitals in the earlier regulations for those provisions for which no recital is currently given.

7.3.1 Confidentiality (Article 2(1(1))

The first item on the white list is confidentiality:

(1) an obligation on the licensee not to divulge the know-how communicated by the licensor; the licensee may be held to this obligation after the agreement has expired;

This provision is identical to article 2(1)(7) of the patent regulation and article 2(1)(1) of the know-how regulation and is not affected by the expiry of any of the patents or the termination of the territorial protection permitted. 'Know-how' is defined by article 10(1) to (4) of the technology transfer regulation, considered at 6.1.1 to 6.1.1.3.

---

[10] [1988] OJ L309/34, [1989] 4 CMLR 535, [1988] 1 CEC 2254, described at 3.1.5 above.
[11] [1989] OJ C188/2, [1990] 4 CMLR 166.
The overlapping products formed only a small part of the entire range. The reason for the reciprocal grant of exclusive distribution rights was that each wanted to get into the other market, not to keep the other out of its own. This was not an example of the application of the *de minimis* notice: together the parties supplied some 20% of the multimeters in the common market, but the market was competitive.
[12] *Nineteenth Report on Competition Policy* (Brussels: EC Commission, 1990), point 47.
[13] It is thought that where the final version of the reg. does not contradict the recitals of the earlier regs., they may be helpful in explaining the Commission's attitude and, doubtless, they should be cited in the Community and national courts.

The confidentiality of marketing and other non-technical secret information is not expressly whitelisted. Nevertheless, the list is expressed in recital 18 and by the introductory words of article 2(1) not to be exhaustive, and it is thought that a provision preserving any confidential information is outside the prohibition of article 85(1).[14] Article 3(2)(a) of the franchising regulation, regulation 4087/88, whitelists unconditionally an obligation not to disclose to third parties the know-how provided by the franchisor, both during and after the termination of the agreement. In that regulation 'know-how' is defined in article 1(3)(f) to mean 'a package of non-patented practical information, resulting from experience and testing by the franchisor'. If its confidentiality could not be preserved, the information might well not be communicated. Moreover, not only the licensee but also some of its employees may be required to agree not to divulge the information. It is thought that a restriction on anyone disclosing any kind of confidential information is outside the prohibition of article 85(1).

The Commission has habitually cleared obligations not to disclose secret technical information in its decisions on exclusive licensing, such as *Burroughs*,[15] *Davidson*,[16] and *Kabelmetal*,[17] and, later, in *Rich Products/Jus-Rol*,[18] and *Delta Cheimie*,[19] at least when the obligation ends once the know-how enters into the public domain. It is sometimes difficult to tell

---

[14] In *Pronuptia de Paris GmbH* v. *Pronuptia de Paris Irmgard Schillgalis* (161/84), [1986] ECR 353, [1986] 1 CMLR 414, CMR 14,245, described at 1.4.1 above, the Court ruled that many obligations imposed on a franchisee were outside art. 85(1). One of the grounds was:
'16. First, the franchisor must be able to communicate his know-how to the franchisees and provide them with the necessary assistance in order to enable them to apply his methods, without running the risk that know-how and assistance might benefit competitors, even indirectly. It follows that provisions which are essential in order to avoid that risk do not constitute restrictions on competition for the purposes of Article 85(1).'
See also the Commission's decisions mentioned in the next para.

[15] *Burroughs/Delpanque* and *Burroughs/Geha* [1972] JO L13/50 and L13/53, [1972] CMLR D67, CMR 9485.

[16] [1972] JO L143/31, [1972] CMLR D52, CMR 9512.

[17] [1975] OJ L222/34, [1975] 2 CMLR D40, CMR 9761.

[18] [1988] OJ L69/21, [1988] 4 CMLR 527, CMR 10,956. The Commission said:
'(32) the licensee's obligation not to disclose, throughout the term of the agreement or subsequently, the confidential information supplied by the licensor must be deemed not to be covered by Article 85(1). The commercial value of know-how, which is a body of technical knowledge not protected by the legal provisions governing industrial property, is dependent on its secrecy. The obligation imposed on the licensee not to disclose the know-how is a necessary condition for maintaining such value and, consquently, for enabling its owner to grant it to other undertakings. This obligation may be maintained after the termination of the licensing agreement as long as the know-how has not come into the public domain. This latter requirement is observed by the agreement notified.'
This argument applies equally to commercial know-how.

[19] N. 10 above, para. 32.

exactly when this happens, as the information slowly spreads through the trade.

It is thought that where the know-how has entered into the public domain otherwise than through the fault of the licensee, it would be unwise to enforce an obligation not to disclose, as the licensee might complain to the Commission, causing the licensor considerable trouble. Where, however, only a limited number of firms have discovered the know-how, it may still be worth preserving it from further dissemination.

The Commission's case law is confirmed by the Court's judgment in *AKZO*,[20] where it held that the Commission's duty not to disclose business secrets overrides its duty to give a complainant a fair hearing by disclosing documents that may be confidential. One would think that this principle should apply *a fortiori* to clear a contractual provision to maintain confidentiality.

As we saw at 6.1.1.1, secrecy is not an absolute concept. Nor is confidentiality. The Commission's reasoning in *Rich Products/Jus-Rol* applies to any information that is sufficiently secret to retain some value.

How far may a licensor require its licensee to buy secret ingredients only from a specified source so as to preserve their confidentiality? In *Campari*,[21] the Commission cleared a requirement that Campari's licensees should obtain the secret herbs exclusively from it in order to preserve the quality of the product. It must also have helped to preserve the secrecy of their formula. Merely to require the licensee to ensure that any employees learning the trade secrets should promise not to disclose them or work for a competing firm may be less effective. A tying provision such as that in *Campari* might also come within article 2(1)(5), as both the obligation and the product would be needed for a technically satisfactory exploitation of the licensed invention.[22] The Commission's decision in *Rich Products/Jus-Rol*[23] on this point was very similar to that in *Campari*. The obligation to buy the secret pre-mix from the licensor was cleared because the parties agreed that it was needed for a satisfactory exploitation of the know-how.

7.3.2 Sub-licensing (Article 2(1)(2))

Article 2(1) also permits

> (2) an obligation on the licensee not to grant sub-licences or assign the licence;

---

[20] *AKZO Chemie BV and AKZO Chemie UK Ltd* v. *Commission* (53/85), [1986] ECR 1965, [1987] 1 CMLR 231, CMR 14,318.
[21] *Re the Agreement of David Campari Milano SpA* (78/253/EEC), [1978] OJ L70/69, [1978] 2 CMLR 397, CMR 10035, 3.1.2 above.
[22] Analysed at 7.3.5–7.3.5.2 below.   [23] N. 18 above.

This corresponds precisely to article 2(1)(5) of the patent regulation and article 2(1)(2) of the know-how regulation. A restriction on sub-licensing without consent was cleared in *Davidson Rubber*[24] on the ground that otherwise the licensor would not be able to protect the confidentiality of its secret information. The order of the white list, where this now follows immediately the obligation of confidentiality, seems to confirm this reasoning.

In *Suralmo*,[25] however, the Commission decided informally that permission to sub-license for civilian use, but only with consent for military use, infringed article 85(1). It closed the file only when the requirement of consent for military use within the common market was dropped. It may be argued that such a partial restraint on sub-licensing could not be justified on grounds of the need to preserve confidentiality, and Suralmo allocated customers within a single field of use.[26] In this case it would not now be exempted by article 2(3) as an obligation of the same type but more limited scope. I regret the absence of recitals stating the purpose of the listed provisions, which might have enabled us to advise clearly on the validity of more limited clauses.

In *Delta Chemie/DDD*,[27] the Commission cleared a prohibition on granting sub-licences without consent on the quite different basis that the owner of the know-how had the sole right to decide whether or not a third party should be allowed to use it. This may be another example of the Commission accepting the theory of a limited licence.

7.3.3 Post-term Use Ban (Recital 20 and Article 2(1)(3))

Recital 20 explains:

> (20) The obligations on the licensee to cease using the licensed technology after the termination of the agreement (Article 2(1)(3)) . . . do not generally restrict competition. The post-term use ban may be regarded as a normal feature of licensing, as otherwise the licensor would be forced to transfer his know-how or patents in perpetuity.

This is implemented first by article 2(1)(3):

> (3) an obligation on the licensee not to exploit the licensed know-how or patents after termination of the agreement in so far and as long as the know-how is still secret or the patents are still in force;

---

[24] N. 16 above.

[25] *Ninth Report on Competition Policy* (Brussels: EC Commission, 1980), 72. The exception from the Treaty for military supplies provided for by art. 223 is far narrower than is generally realized. It is only state action in relation to military supplies that may be excepted. *Suralmo* concerned an agreement between firms in the private sector.

[26] See art. 3(4), 7.3.8.2 below.   [27] N. 10 above, at para. 36.

It is doubted whether the express reference in sub-paragraph (3) to the know-how being secret impliedly dispenses with its being substantial. Any questions about this are unlikely to be important, since the remaining know-how would have to be trivial not to qualify.

Where it is expensive for the licensee to set up production, it may not be possible to persuade anyone to take a licence for a short term, but where existing plant can be adapted cheaply as between different uses, an owner of technology that has not got sufficient resources itself to manufacture throughout the common market may prefer to exploit its idea by licensing initially for a limited period.[28]

The legality of a ban on using the technology after the licence has expired used to be controversial. Since there is no intellectual property right, the licensee is restricted by contract from doing what is permitted by the general law.[29] This formalistic view gave way to the concept that one should be able to 'lease' as well as 'sell' the technology. Recital 20[30] makes it clear that the Commission has changed its view:

> The post-term use ban may be regarded as a normal feature of licensing, as otherwise the licensor would be forced to transfer his know-how or patents in perpetuity.

This is yet another example of the Commission analysing the licence as being limited.[31] It is limited in time, but the Commission does not pursue

---

[28] In *Velcro/Aplix*, n. 7 above, Velcro did not start to manufacture until the basic patents had expired. Presumably, since the firm was tiny at the time the licence was first granted, it could not afford to set up production until it had collected significant royalties from its licensees.

[29] See the cases described at 3.1.4, n. 30, above. Such a clause was condemned in *Cartoux/ Terrapin*, an informal decision published in the *Tenth Report on Competition Policy* (Brussels: EC Commission, 1981), point 129, and described at 3.1.4, n. 30 above. A ban on using the technology after the end of the term was blacklisted by art. 3(10) in the first published draft of the patent reg. [1979] OJ C58/12, [1979] 1 CMLR 478, but did not appear in any of the lists of the final version. Consequently, such a clause was thought to fall within the opposition procedure. The Commission completed its U-turn and whitelisted the clause in art. 2(1)(3) of the know-how reg.

In the 1970s, the Commission took the formalistic view that any restrictions on conduct that were not within the scope of the patent infringed art. 85(1). See **James Venit**, 'EEC Patent Licensing Revisited: The Commission's Patent License Reg.' (1985) XXX *Antitrust Bulletin* 457, 475:

> 'In essence, the Commission appeared to have come to the conclusion that the only right inherent in know-how is its secrecy, but that know-how cannot, properly speaking, be licensed, with the result that any restriction as to its field of application or the territories in which it can be used will be caught by art. 85(1).'

Compare the passage quoted from the AG's opinion in *Windsurfing International Inc.* v. *Commission* (193/83), [1986] ECR 611, [1986] 3 CMLR 489, CMR 14,271, at 6.6.4 above, text to n. 44 ff.

The Commission's views have changed dramatically since then. It is now thinking in terms of technology licensing and the need to induce risky investment.

[30] Like recital 14 of the know-how reg.   [31] See 3.2.5 above.

this theory to the limit. Territorial limitations are still treated as restrictions on exploitation and exempted by article 1(1)(1) to 1(1)(6).

Article 3(1) of the know-how regulation limited the post-term use ban to the period during which the know-how remained secret, unless it became public through the unlawful action of the licensee.[32] The limitation to the period during which the know-how remains secret or the patent remains in force is now contained in article 2(1)(3). So it is prudent expressly to limit a ban on using the know-how after termination of the licence to the period when substantial know-how remains secret.[33] A licensor may look at the common law on restraint of trade, although it is no part of Community law, and provide for the restriction on use to end once the information becomes generally available and in widespread use, and require that the licensee should establish to the reasonable satisfaction of the licensor that the know-how has entered the public domain.

If licences are granted to A, B, and C, each to exploit the technology only for four years, and A discloses the technology publicly, A can be held to the ban—he cannot profit from his own wrong—but B and C must be permitted to continue to exploit the technology, even if they obtained a head start during the period of their licences. This possibility may well discourage each licensee from disclosing the information. Care should also be taken to inform each licensee before the contract is concluded of the likely loss if the licensee discloses the technology wrongfully and to record such a warning, in order to prevent much of the damage being too remote.

The provisions on feed-back limit the value of this item less than under the know-how regulation. See 7.3.4 to 7.3.4.4 below.

7.3.4 Feed- and Grant-back (Recital 20 and Articles 2(1)(4) and 3(6))

Feed- and grant-back provisions may be necessary to persuade the holder of technology to grant a licence. The holder of the technology may not be prepared to create a competitor that may improve the technology and keep the improvements to itself. This argument is particularly strong when the licensor agrees to pass on to its licensee any improvements that it may make or which it is authorized to pass on. Most licences are for the best

---

[32] Compare *Rich Products/Jus-Rol*, cited n. 18 above, where a ban on using the technology for 10 years after termination of the agreement was cleared, but the Commission stated that the ban would become anti-competitive if the know-how became freely accessible through no fault of the licensee, or if the licensee obtained it from a third party.

[33] Where a licensed patent remains in force, but all the know-how has clearly entered into the public domain, it would be dangerous to try to restrain the licensee from continuing to use the know-how. Even if the ban be valid, the Commission might be persuaded to threaten to revoke the exemption. Moreover, the Community Courts and Commission do not always construe Community legislation by reference to its literal meaning.

way of accomplishing a result and include an obligation for the licensor to pass on improvements that he makes or is authorized to impart.[34]

Obligations to feed or grant back improvements and new applications may lead to better use being made of innovations that are made, but the need to share them reduces the incentives for licensees to invest in innovation. The Commission has compromised in all three regulations for technology transfer. It will permit such clauses, if they are reciprocal and non-exclusive. This enables the licensee to use its own improvements and if there are people whom it may license without disclosing the licensor's secret know-how, it will retain some incentive to invest in innovation.

In the know-how regulation the Commission also insisted that the licensee must be given a chance to negotiate a further licence at the termination of the original licence if its improvements have been useful. The Commission has also tried to preserve the right of the licensee to license its own improvements to third parties, where this will not disclose the licensor's know-how. These provisions have been repeated in the technology transfer regulation. See 7.3.4.3 below.

In the United States, a rule of reason is applied to licences of intellectual property rights, and the market power of the licensor will be an important factor in the analysis by the Agencies. According to the *Antitrust Guidelines for the Licensing of Intellectual Property*:[35]

5.6 Grantbacks

A grant back is an arrangement under which a licensee agrees to extend to the licensor of intellectual property the right to use the licensee's improvements to the licensed technology. Grantbacks can have pro-competitive effects, especially if they are non-exclusive. Such arrangements provide a means for the licensee and the licensor to share risks and reward the licensor for making possible further innovation based on or informed by the licensed technology, and both promote innovation in the first place and promote the subsequent licensing of the results of the innovation. Grant-backs may adversely affect competition, however, if they substantially reduce the licensee's incentives to engage in research and development and thereby limit rivalry in innovation markets.

A non-exclusive grantback allows the licensee to practice its technology and license it to others. Such a grantback provision may be necessary to ensure that the licensor is not prevented from effectively competing because it is denied access to improvements developed with the aid of its own technology. Compared with an exclusive grantback, a non-exclusive grant-

---

[34] This may not be the case when the licence is granted as part of a settlement of patent infringement litigation, or where an academic institution has lost interest in an idea soon after applying for the basic patents.
[35] Issued by the US Department of Justice and the Federal Trade Commission, 6 Apr. 1995, reproduced in CCH, 4 *Trade Regulation Reporter*, ¶13,132, (1995) 7 *EIPR* Supp. 3, and considered generally at 1.4.3 above.

back, which leaves the licensee free to license improvements technology to others, is less likely to have anticompetitive effects.

The agencies will evaluate a grantback provision under the rule of reason, . . . considering its likely effects in the light of the overall structure of the licensing arrangement and conditions in the relevant markets. An important factor in the agencies' analysis of a grantback will be whether the licensor has market power in a relevant technology or innovation market. If the Agencies determine that a particular grantback provision is likely to reduce significantly licensees' incentives to invest in improving the licensed technology, the Agencies will consider the extent to which the grantback provision has offsetting procompetitive effects, such as (1) promoting dissemination of licensees' improvements to the technology, (2) increasing the licensors' incentives to disseminate the licensed technology, or (3) otherwise increasing competition and output in a relevant technology or innovation market. See section 7.2. In addition the Agencies will consider the extent to which grantback provisions in the relevant markets generally increase licensors' incentives to innovate in the first place.

In a group exemption, the EC Commission cannot assess the licensor's market power in an innovation market, or the effects of a particular grantback provision. Instead, it has developed automatic rules. In recital 20 of the technology transfer regulation, the Commission explains its compromise:

(20) The obligations on the licensee . . . to make improvements available to the licensor (article 2(1)(4)) do not generally restrict competition. . . . Undertakings by the licensee to grant back to the licensor a licence for improvements to the licensed know-how and/or patents are generally not restrictive of competition if the licensee is entitled by the contract to share in future experience and inventions made by the licensor. On the other hand, a restrictive effect on competition arises where the agreement obliges the licensee to assign to the licensor rights to improvements of the originally licensed technology that he himself has brought about.

7.3.4.1 Weak Feedback is Whitelisted (Article 2(1)(4))

Article 2(1)(4) permits:

(4) an obligation on the licensee to grant to the licensor a licence in respect of his own improvements to or his new applications of the licensed technology, provided:
—that, in the case of severable improvements, such a licence is not exclusive, so that the licensee is free to use his own improvements or to license them to third parties, in so far as that does not involve disclosure of the know-how communicated by the licensor that is still secret,
—and that the licensor undertakes to grant an exclusive or non-exclusive licence of his own improvements to the licensee;

If the licensor limits the use of technology to a term, it must watch the feed-back provisions and must either renew the initial licence or lose the benefit of the licensee's improvements, past or future.[36] With careful drafting of the initial agreement, the licensor may arrange to decide whether to grant a further licence only at the end of the initial term, by which time he will have a better idea of the usefulness of the licensee's improvements.

7.3.4.2 Reciprocal Licences of Feedback Seldom Identical

As is the case under the earlier regulations, the licences the parties grant each other need not be identical. Where the original licence is exclusive, the technology passed on to the licensee is likely also to be licensed exclusively, but the feedback will be non-exclusive unless it cannot be used without breaking the confidentiality provisions.

Normally, the licence and feedback will apply to different territories. Indeed, the licensor may well want a worldwide licence for the licensee's improvements, with a right to sub-license. The licensor may be permitted to sub-license only to those licensees who have agreed to let him have their improvements and to pass them on to other licensees. Such an agreement may be pro-competitive in spreading any improvements that occur spontaneously around the network, but may remove most of the incentive for licensees to invest in innovation and so be restrictive where innovation requires significant investment, especially when research and development is virtually the only kind of competition possible. If the licensor has undertaken to provide the improvements for the royalty fixed under the original licence, one licensee will not pay royalties to an innovating licensee for the same development. Occasionally, there will be a market between the licensees for granting licences for improvements and new applications and the innovating licensee will be able to negotiate a royalty. It is thought that any of these arrangements qualify. Only if the know-how is so closely connected to that of the licensor that it is useless without disclosing the latter, can the feed back clause be exclusive.[37]

7.3.4.3 The Provisos to Article 2(1)(4)

The first proviso prevents the licensor from requiring feed- or grant-back and restraining the licensee from licensing its own improvements to third parties if they can be used separately from the licensor's technology, and there is no need to disclose the latter.

---

[36] Recital 20 of the technology transfer reg.
[37] Art. 3(6), discussed at 7.3.4.4 below.

—that, in the case of severable improvements, such a licence is not exclusive, so that the licensee is free to use his own improvements or to license them to third parties, in so far as that does not involve disclosure of the know-how communicated by the licensor that is still secret,

Under the equivalent provision in the know-how regulation, the licensor might require to be informed before the licensee granted a licence of its own improvements, but could restrain such licensing only if justified on grounds of preserving the confidentiality of the know-how. These provisions have been deleted in the technology transfer regulation but, in view of article 2(1)(1) and (2), which protects the preservation of confidentiality, it is thought that the provisions about the burden of proof were excluded only to simplify the regulation, and that the legal position has not altered. There is probably no need to use the opposition procedure.

By virtue of the second proviso, for article 2(1)(4) to apply, the licensor must accept an obligation to communicate its own improvements to the licensee:

—and that the licensor undertakes to grant an exclusive or non-exclusive licence of his own improvements to the licensee;

According to recital 20, the licensor's right to use the licensee's improvements that are not severable from the licensor's technology must not extend beyond the date on which the licensee's right to use the licensor's improvements comes to an end. This enables a licensee who develops useful know-how to renegotiate a further licence in return for a cross-licence.

The provisos reduce the value of a post-term ban on use of the licensed technology, when the licensee's improvements are useful,[38] but is no more restrictive than the patent licensing regulation, which does not whitelist restrictions on using the technology after the term expires.

If the provisos are not satisfied, but some other inducement for licensees to invest in innovation is provided, such as reasonable payment to be assessed, in the case of dispute, by arbitration, the question arises whether the post term use ban escapes the prohibition of article 85(1) on the basis that the list is not exhaustive. This might just be argued.[39]

Alternatively, the parties might consider notifying the agreement under the opposition procedure, unless article 3(2) or (6) applies. There will be

---

[38] The Commission's reasoning is spelled out in *Delta Chemie/DDD*, n. 10 above, at para. 33.

[39] See, however, the Commission's decision in *Velcro/Aplix*, n. 7 above, where such a provision was treated as infringing art. 85(1) because it was blacklisted. The Commission's views have changed significantly since that date on many points, so I hope the precedent is no longer useful.

unlisted provisions that the Commission may think restrict competition: the ban on using the technology after the term has expired coupled with slightly too strong a feedback clause. Article 3(6) does not circumscribe article 2(1)(4).

The licensor may reserve the option referred to in the second proviso to article 2(1)(4) in the original licence and decide which alternative to take on termination, when he can evaluate the licensee's improvements.

7.3.4.4 Strong Feedback Clauses are Blacklisted (Article 3(6))

Strong feed back provisions are blacklisted. Article 3(6) prevents the application of the regulation where:

> (6) the licensee is obliged to assign in whole or in part to the licensor rights to improvements to or new applications of the licensed technology;

Article 3(6) is similar to article 3(2)(a) of the know-how regulation and article 3(8) of the patent regulation: an obligation to assign improvements to or new applications of the technology in whole or in part prevents the application of the regulation. The reason given in recital 24 of the patent regulation for blacklisting strong grant-back clauses is that they put the licensee at a competitive disadvantage. The reason given in earlier individual decisions is that such a feedback clause removes the incentive for the licensee to invest in innovation. More recently, *Rich Products/Jus-Rol* is based on the incentive to innovate and *Delta Chemie*[40] on both theories. The first part of recital 14 of the know-how regulation also seems to be based on the incentive to innovate, although the second paragraph is based on the ability of the licensee to use its own innovations.

The question arises whether, if these reasons do not apply, for instance because there is provision for fixing a reasonable royalty to be paid for improvements, the black list applies. Even in an individual decision in *Velcro/Aplix*[41] this did not prevent the clause being condemned as a misuse of patent without any analysis. Although that attitude of the Commission towards blacklisted provisions when it considers individual agreements may have changed,[42] article 3(6) seems to me clearly to prevent the automatic application of the group exemption.

The other provisions in article 3(2) of the know-how regulation have been deleted. They circumscribed the provisos in article 2(1)(4) discussed at 7.3.4.3 above. The question arises whether such obligations are now considered to infringe article 85(1) and to require notification under the

---

[40] Nn. 18 and 10 above respectively.
[41] Cited at n. 7 above and discussed at the text thereto.
[42] See 7.2 above.

opposition procedure. It might be argued that where the licensor has little market power they would not be treated as restrictions of competition.

I see no reason why the Commission should be concerned to protect licensees from making bad bargains. Usually, it is a licensor which lacks manufacturing capacity that grants exclusive licences. Such a licensor may well be smaller than the licensee. On the other hand, the provisos to article 2(1)(4) do preserve the incentive for the licensee to invest in innovation.

7.3.5 Minimum Quality and Tying (Articles 2(1)(5) and 4(2)(a))

Tying is one of the examples given in articles 85 and 86 of illegal conduct, and some people believe that tying enables a firm with market power to extend that power from the tying to the tied product. This is denied by most economists on the ground that there is only one monopoly profit to be made from the tying product and that, if the licensor requires licensees to buy the tied product from him, they will pay less for the tying product. Chicago economists consider, therefore, that tying is more likely to be adopted to achieve efficiencies, most often that it is used to monitor royalties: one way of discovering how much use is being made of the tying product is to require the buyer or licensee to buy some consumable from the supplier or licensor. The large user is likely to be prepared to pay more for the tying product than would a smaller user.

A second justification that has been accepted by the Commission and the Court is that the tied product may be necessary for a technically adequate exploitation of the technology. There are many other justifications for tying.[43]

The US guidelines on intellectual property licensing[44] look to the structure of the market as well as to the justifications when assessing tying.

5.3 Tying arrangements

A 'tying' or 'tie in' or 'tied sale' arrangement has been defined as 'an agreement by a party to sell one product . . . on the condition that the buyer

---

[43] The following is a non-exhaustive list of reasons for tying that have not been accepted by Commission or Court.

1 In the US, in *Jerrold Electronics* v. *US*, 187 F. Supp. 545 (ED Pa. 1960), affirmed *per curiam*, 365 US 567 (1961), it was recognized that when complex new equipment, such as cable television, is first introduced it may be necessary to tie all the parts and service in order to prevent the reputation of the products being ruined by faulty performance.

2. If charges for the tying product are regulated, the supplier may prefer to take its profit in another market.

3. Where customers want a package, can it be called a tie? Is a vehicle a tie of its parts? Are laces tied to shoes? They could be sold separately with a notice attached to the shoe as to the length of the appropriate laces.

4. Selling in bulk may economize in grading and in transport costs.

5. A tie may be one way a firm can give a secret discount in an oligopolistic market.

[44] N. 35 above.

also purchases a different (or tied) product, or at least agrees that he will not purchase that [tied] product from any other supplier.' *Eastman Kodak Co.* v. *Image Technical Services Inc.*, 112 S. Ct. 2072, 2079 (1992). Conditioning the· ability of a licensor to license one or more items of intellectual property on the licensee's purchase of another item of intellectual property or a good or a service has been held in some cases to constitute illegal tying. Although tying arrangements may result in anticompetitive effects, such arrangements can also result in significant efficiencies and procompetitive benefits. In the exercise of their prosecutorial discretion, the Agencies will consider both the anticompetitive effects and the efficiencies attributable to a tie-in. The Agencies would be likely to challenge a tying arrangement if: (1) the seller has market power in the tying product, (2) the arrangement has an adverse effect on competition in the relevant market for the tied product, and (3) efficiency justifications for the arrangement do not outweigh the anti-competitive effects. The Agencies will not presume that a patent, copyright or trade secret necessarily confers market power on the owner.

Package licensing—the licensing of multiple items of intellectual property in a single license or in a group of related licences—may be a form of tying arrangement if the licensing of one product is conditioned upon the acceptance of a license of another separate product. Package licensing can be efficiency enhancing under some circumstances. When multiple licenses are needed to use any single item of intellectual property, for example, a package license may promote such efficiencies. If a package license constitutes a tying arrangement, the Agencies will evaluate its competitive effects under the same principles they apply to other tying arrangements.

A European group exemption, however, cannot rely on such a market analysis.

7.3.5.1 Minimum Quality and Tying are Whitelisted When Necessary for the Technically Satisfactory Exploitation of the Technology, or Respected by the Licensor and Other Licensees (Article 2(1)(5))

Minimum quality specifications observed by licensor and other licensees and an obligation to procure goods or services from a specified source are whitelisted by article 2(1)(5) when necessary for a technically satisfactory exploitation of the licensed technology.

The quality specifications must be actually observed by licensor and licensees, and not merely 'applicable' as in article 4(2)(a). The French text draws a similar distinction. Where the quality specifications are not observed by some licensees, it is thought that there should be no problem under the opposition procedure, unless it can be established that the licensor intended there to be different standards in fact.

In *Rich Products/Jus-Rol*,[45] the Commission cleared an obligation to buy

---

[45] N. 18 above, at para. 37.

from the licensor a pre-mix whose composition was secret. Emphasizing the practical nature of this arrangement, the Commission observed that the parties took the view that it was necessary to a technically satisfactory exploitation of the licensed technology and accepted it. The licensor was not required to disclose the composition of the pre-mix to its licensees. The Commission did not expressly consider whether, if the pre-mix really is necessary for satisfactory exploitation, there would be any need to impose the tie. Since the licensee would take the tied product anyway, there may have been some other reason for the tie; the licensor might have wished to ensure that the licensee maintained standards; alternatively, it may have been of no importance.

In *Campari*,[46] the requirement to obtain the secret herbs from Campari was justified by the need to ensure a uniform flavour and quality of the product sold under the Campari mark, although the widely differing alcoholic strengths and the use of local wines may have resulted in different tastes. The Commission considered that the tie was justified by the need for uniformity of a franchised product. That justification was not suggested in *Rich Products/Jus-Rol*. Other justifications often given for a tie are that the use of the product bought in measures the use made of the technology, so royalties can be collected more easily.[47] and that it enables a regulated monopolist to take his profit outside the regulated market.

Article 2(1)(5) whitelists:

(5) an obligation on the licensee to observe minimum quality specifications, including technical specifications, for the licensed product or to procure goods or services from the licensor or from an undertaking designated by the licensor, in so far as these quality specifications, products or services are necessary for:
  (a) a technically proper exploitation of the licensed technology; or
  (b) ensuring that the product of the licensee conforms to the minimum quality specifications that are applicable to the licensor and other licensees;

The know-how regulation added: 'and to allow the licensor to carry out related checks'. It is thought that this has been omitted solely for the sake of simplicity and is not restrictive of competition.[48]

---

[46] N. 21 above.
[47] See remarks of Abbott B. Lipsky, Jr., the Deputy Attorney General, Antitrust Div., before the ABA Antitrust Section, 1981, conveniently reproduced in S. C. Oppenheim, G. E. Weston and J. T. McCarthy, 1985 *Supplement to Federal Antitrust Law* (4th edn., St Paul, Minn.: West Publishing) 129.
See also the justifications listed at n. 43 above.
[48] In *Raymond/Nagoya* [1972] JO 143/39, [1972] CMLR D45, 50, CMR 9513, the Commission cleared a provision for quality inspections. Since, however, the licence was for

Paragraph (b) was introduced by the know-how regulation and goes further than the patent regulation, but some licensors may want to go further and specify different quality standards for different geographic or product markets. The question arises as to the position when some of the standards are stricter than others, or differ.[49] The clause may be exempted by the regulation on the ground that the list in article 2 is not exhaustive. Alternatively, since the whitelisted clause is no longer circumscribed by the black list,[50] there is a possibility of using the opposition procedure.

The question also arises whether the licensor must enforce the quality standards required of other licensees if quality standards are to be enforced against one of them. Article 2(1)(5)(b) speaks of 'minimum quality specifications that are applicable to licensor and other licensees'. This is simpler than the provision in the know-how regulation, which required the minimum quality specifications to be respected by licensor and other licensees.

7.3.5.2 *In Other Circumstances Unwanted Ties and Quality Specifications are Subject to the Opposition Procedure (Article 4(2)(a))*

The whitelisted provision is circumscribed by article 4(2)(a) in the technology transfer regulation and subjected to the opposition procedure, not blacklisted as in the know-how regulation. It applies not only to unwanted goods and services, but also to unwanted quality specifications, unless they are necessary for ensuring that the licensee conforms to quality standards that are required for a technically satisfactory exploitation of the technology licensed or are respected by the licensor and other licensees.

The opposition procedure is described in Chapter 8 below. Where there is a pure patent, know-how, or mixed licensing agreement within the meaning of article 1(1) and no blacklisted provisions, but some provision that infringes article 85(1) and is not exempted, the parties may notify the Commission, and if the Commission does not oppose the exemption within four months, the whole agreement will be exempt under the regulation.

Article 4(2) lists two specific kinds of provisions to which the opposition procedure applies, probably, even if they do not happen to infringe article 85(1). The first is:

---

manufacture in Japan, the agreement had little effect in the common market and the precedent is, therefore, weak.

In *Burroughs*, n. 15 above, the Commission cleared a duty to mark on the ground that it facilitated checks on quality.

[49] See 7.3.5.2, at the end.

[50] Art. 3(3) in the know-how reg. has been moved to the opposition procedure, art. 4(2)(a) of the technology transfer reg.

(a) the licensee is obliged at the time the agreement is entered into to accept quality specifications or further licences or to procure goods or services which are not necessary for a technically satisfactory exploitation of the licensed technology or for ensuring that the production of the licensee conforms to the quality standards that are respected by the licensor and other licensees;

Provided the standards are consistently required, the licensor may select the level of quality he wants. There is nothing in the regulation requiring only a middling quality to be specified. In *Rich Products/Jus-Rol*,[51] an exemption was granted although it was possible to make frozen yeast dough without using Rich's pre-mix, since the latter led to more even results.

There are several old cases where the Commission condemned tying, but some officials seem now to accept that tying may be a good way of monitoring royalties. There must be some disagreement, because the Commission has retained control over such agreements by providing that they may fall under the group exemption only if notified to the Commission under article 4, and if the Commission does not oppose the exemption within four months.

When unwanted ties were blacklisted, it was very important to try to find reasons for not calling them unwanted ties. The reasons are now less acute since they are subject only to the opposition procedure. It suffices to notify and justify them economically.

Among other clauses, the Commission disapproved of tying in *Vaessen/Moris*.[52] M. Moris granted a royalty-free licence of his patented device for filling *saucissons de Boulogne*, on condition that the licensees bought their skins from him. These he supplied bearing the number of the patent. If his device was used, only a single skin was necessary, whereas the traditional method of filling required a double skin. If he found sausages in a single skin not bearing the patent number, M. Moris would know that someone was cheating and he could investigate. The tie was a good way of measuring the use of the device. Nevertheless, the Commission condemned it as foreclosing other makers of skins whose prices were lower. Of course third parties' prices were lower: they could claim no reward to induce them to make the patented invention available.

Would such an agreement now come within the opposition procedure? If M. Moris would have supplied the patented device royalty free without the tie, no doubt the licensee would have been happy. In that sense it did not want the tie. If, however, the choice was between paying a higher royalty

---

[51] N. 18 above.
[52] [1979] OJ L19/32, [1979] 1 CMLR 511, CMR 10,107; criticized by **Lucio Zanon**, 'Current Survey, "Ties in Patent Licensing Agreements" ', (1980) 5 *ELRev.* 391.

(to allow for inefficiency in collecting royalties) or paying a little extra for each skin, an honest licensee might well want the tie. The concept of particular clauses that the licensee is obliged to accept is a difficult one to apply. Usually each party insists on certain clauses and, looked at in isolation, the other party does not want them. It is hoped that this is not what is meant by the phrase, and that one should look to the realistic alternatives at the time the licence was being negotiated. At least, it is made clear by article 4(2)(a) that it is *ex ante* that one decides whether the tie or specification is wanted. A licensee who expects to want to have a chance to negotiate may consider recording in his file that he does not want the tie, or specification, while the licensor may negotiate to have an acknowledgement in the contract that the licensee does want it recited in the contract.

In *Velcro/Aplix*,[53] the Commission objected to the requirement that licensees should buy their equipment from a particular supplier. This provision also made commercial sense: the inventor seems to have had few resources other than its invention, at least until the royalties enabled it to invest in production facilities of its own, and it persuaded another firm to develop the necessary equipment in return for a promise that its licensees would buy only from it. Without such a tie, it might well not have been possible to get the important innovation into production.

The decision related only to the period after the basic patents had expired. By that time it might have been possible to find alternative sources of machinery, since the technology had proved very successful. By that time new licensees would not want the tie, since the investment of the equipment builder had already been made, although without the expectation of the tie it might not have been and the technology might never have been exploited. The Commission's decision is another example of the Commission analysing *ex post*.[54]

Sometimes, the holder of know-how does not want to pass on the secret recipe in a know-how licence and, instead, provides a secret bundle of herbs, pre-mix, or whatever. The regulation speaks of the specification or products being necessary and, if the recipe is not known, then it may be necessary to obtain the products made by its use, without the licensor having to establish that there is no other satisfactory specification or source. This is confirmed by the decision in *Rich Products/Jus-Rol*. The pre-mix was secret, and the Commission permitted the licensee to be required to obtain it from the licensor without requiring the parties to establish that the licensor would not have been prepared to disclose the recipe. Where the tied product is not confidential, I would hesitate to rely on the literal reading that it is the product that must be necessary rather

---

[53] N. 7 above.   [54] See 1.3.1 above.

than the tie when a satisfactory product could have been obtained elsewhere. That seems to make no sense from a policy viewpoint.

Under section 44(1) of the UK Patents Act 1977, any condition or term of a contract to supply a patented product or to license a patented process is void if the supplier or licensor requires anything else to be acquired from him or from a nominee. As long as the term is in existence, it constitutes a defence to any infringement action by the patentee. The law is harsh, but has been construed narrowly by the courts. Higher royalties may be charged if a specified product is not bought from the licensor or his nominee, and this has been held to be valid.[55] That precedent is not likely to be followed under EC law, which looks more to the spirit of the legislation than to its wording. It is irrelevant under the UK Patents Act whether the tied product is technically necessary.

Article 1(1)(8) exempts use licences and, at 6.6.4 above, I suggested that one might look on them as a kind of tie. The licensee may use the licensed product only for its own production of something else—only with the rest of the final product it is producing. That, however, is not subject to the opposition procedure even if the licensee would have preferred to be granted a less narrowly limited licence.

In several decisions the Commission has been prepared to clear provision for checks to ensure that obligations that do not infringe article 85(1) are respected. Could it now be argued that the tie in *Vaessen/Moris* came within the white list because it enabled the licensor to monitor the use of its device? I doubt it. Nevertheless, if notification be made under the opposition procedure, the agreement might well pass through, and a press release be made to provide a precedent for the future.

Sometimes, different classes of user may need different qualities of product. By requiring high quality standards, one may confine a licensee to serving a particular class of client, but this may be done under these provisions only if all licensees are required to adopt the same standards. Such a division of customers may be achieved under article 2(1)(8), circumscribed by article 3(4), which deal with field of use and customer restrictions.[56] They do not, however, provide for differing minimum qualities of product. That would be subject to the opposition procedure.

The final words of articles 2(1)(5)(b) and 4(2)(a) may cause difficulties where the licensor considers that different quality specifications for products bearing its mark are appropriate in different territories because of

---

[55] *Tool Metal* v. *Tungsten Electric* [1955] 1 WLR 761, at 775–9 (WLR), [1955] 2 All ER 657, (1955) 72 RPC 209 (HL) per Lord Oaksey, reading the judgment that Lord Reid would have delivered had he taken the oath by the date of judgment.

See, e.g., **William R. Cornish**, *Intellectual Property: Patents, Copyright, Trademarks and Allied Rights* (2nd edn., London: Sweet & Maxwell, 1989), 196. The 3rd edn. should be out about the same time as this book.　　　　　　　　　　　　　[56] See 7.3.8–7.3.8.2 below.

different requirements. It has been suggested that the minimum quantity or tying should be permitted when necessary to preserve the reputation of the licensor or other licensees, but it would be difficult for such an interpretation to be given to the words actually adopted, except under the opposition procedure. Where machinery is required to meet tighter tolerances in some areas because of the prevailing terrain, the holder might license a different trade mark and state in the contract that the stricter standards are necessary for a technically proper exploitation of the licensed technology to produce machinery for use in the territory. It may be necessary to notify under the opposition procedure when only tastes are different.

7.3.6 Assistance in Discovering Misappropriation—Restrictions on Challenging the Validity of Intellectual Property Rights, on Contesting the Secrecy of Know-how or the Necessity of the Patents (Articles 2(1)(6), (15), and (16) and 4(2)(b))

7.3.6.1 Whitelisted Provision (Article 2(1)(6))

As under the earlier regulations, article 2(1)(6) provides that the licensee may be required to help the licensor discover misappropriation of the know-how or infringement of the licensed patents and take legal action against the wrongdoer. This obligation used to be without prejudice to the licensee's right to contest the secrecy of the know-how, save where he has himself contributed to its disclosure. That proviso has, however, been deleted.

Article 2(1)(6) whitelists:

(6) obligations:

(a) to inform the licensor of misappropriation of the know-how or of infringements of the licensed patents; or

(b) to take or to assist the licensor in taking legal action against such misappropriation or infringements;

This provision must be subject to article 3(3) which prevents the regulation from applying if either or both parties are required to make it difficult for users or resellers to obtain the products from other resellers within the common market. Neither party may be required to exercise intellectual property rights to prevent parallel trade when the rights are exhausted.[57]

In *Decca Navigator Systems*,[58] the Commission took the view that no industrial property rights existed, but added at paragraph 119 that, if any

---

[57] See 2.2.2–2.2.3 above.
[58] [1989] OJ L43/27, [1990] 4 CMLR 627, [1989] 1 CEC 2137, at para. 104.

copyright had existed, an obligation imposed on the licensees to take legal action against unlicensed suppliers would reinforce the effects of an agreement intended to impede the entry of a new competitor and to divide the market. Consequently, the obligation infringed article 85(1). The Commission seems to have changed its mind. In any event, article 2(2) exempts provisions listed in paragraph (1), where they happen to infringe article 85(1), so a requirement to help the licensor to enforce its rights does not prevent the application of the regulation.

7.3.6.2 *Restrictions on Challenging Validity of Patents or Secrecy and Substantial Nature of Know-how are Subject to Opposition Procedure (Article 4(2)(b))*

Article 4(2)(b) of the regulation subjects a licence to the opposition procedure if:

> (b) the licensee is prohibited from contesting the secrecy or the substantiality of the licensed know-how or from challenging the validity of patents licensed within the common market belonging to the licensor or undertakings connected with him.

This item extends to prohibitions on challenging the secrecy, but not the ownership, of the know-how, or the validity, but not the ownership, of any patents licensed within the common market, belonging to the licensor. In *Moosehead/Whitbread*,[59] the Commission cleared restrictions on

---

[59] [1990] OJ L100/32, [1991] 4 CMLR 391, [1990] 1 CEC 2127, at para. 15(4), the Commission stated:
'(a) In general terms, a trade mark non-challenge clause can refer to the ownership and/or validity of the trade mark:
—The ownership of a trade mark may, in particular, be challenged on grounds of the prior use or prior registration of an identical trade mark.
   A clause in an exclusive trade mark licence agreement obliging the licensee not to challenge the ownership of a trade mark, . . . does not constitute a restriction of competition within the meaning of Article 85(1). Whether or not the licensor or licensee has the ownership of the trade mark, the use of it by any other party is prevented in any event, and competition would thus not be affected.
—The validity of a trade mark may be contested on any ground under national law, and in particular on the grounds that it is generic or descriptive in nature. In such an event, should the challenge be upheld, the trade mark may fall within the public domain and may thereafter be used without restriction by the licensee and any other party.
Such a clause may constitute a restriction of competition within the meaning of Article 85(1), because it may contribute to the maintenance of a trade mark that would be an unjustified barrier to entry into a given market.
Moreover, in order for any restriction of competition to fall under Article 85(1), it must be appreciable. The ownership of a trade mark only gives the holder the exclusive

challenging the ownership of a mark on grounds that seem to apply equally to patents: that any foreclosure of third parties is not dependent on who holds the rights. It stated that restrictions on challenging validity might infringe article 85(1) when these have an appreciable effect. It is only the latter that are subjected to the opposition procedure.

Unlike article 3(1) of the patent regulation, article 4(2)(b) does not extend to other commercial property rights, such as trade marks. Nor does it apply to rights outside the common market.

Prohibitions on challenging the validity of patents or the secrecy of know-how used to be blacklisted, but are now only subject to the opposition procedure. This represents a considerable softening of the Commission's opposition. The change is most welcome.

Moreover, when no-challenge clauses were blacklisted, this was subject to a proviso that the licensor might reserve the right to determine the licence in the event of challenge. It used not to be entirely clear whether such a reservation was whitelisted, or to be treated as a restriction of competition subject to the opposition procedure. It is now expressly whitelisted. Article 2(1)(15) provides that:

> (15) a reservation by the licensor of the right to terminate the agreement if the licensee contests the secret or substantial nature of the licensed know-how or challenges the validity of licensed patents within the common market belonging to the licensor or undertakings connected with him;

Article 2(1)(16) provides that:

> (16) a reservation by the licensor of the right to terminate the licence agreement of a patent if the licensee raises the claim that such a patent is not necessary;

The necessity of a patent is relevant to the duration permitted by article 1(4) of territorial restraints in a mixed patent and know-how licence. Only if the patents are necessary may the territorial restraints last beyond the five- and ten-year periods permitted for know-how licences, for the period during which a patent remains valid in both the licensed territory and that protected. Reservation of the right to terminate the licence may well deter such a claim by the licensee.

The know-how may not be useful without the patent licence, but if it is,

---

    right to sell products under that name. Other parties are free to sell the product in question under a different trade mark or trade name. Only where the use of a well-known trade mark would be an important advantage to any company entering or competing in any given market and the absence of which therefore constitutes a significant barrier to entry, would this clause which impedes the licensee to challenge the validity of the trade mark, constitute an appreciable restriction of competition within the meaning of Article 85(1).'

the question arises whether the licensor can reserve the right to terminate the licence of the know-how too. Article 1(4) seems to treat a mixed know-how and patent licence as a single agreement and article 2(1)(14) enables the licensor to reserve the right to terminate the agreement and not merely the patent licence. In face of this slight uncertainty, the English tradition would be to draft the reservation in the terms of the regulation so as to ensure that the agreement is valid, and determine its meaning when the licensor wishes to terminate it. The civil law tradition seems to be to decide what it means at the time the contract is made.

7.3.6.3 Change in Commission's Hostility to No-challenge Clauses

The Commission has been hostile to no-challenge clauses since its decision in *Davidson Rubber*, where such a clause was abrogated before the Commission would exempt even manufacturing exclusivity. The Commission stated that the licensee had a greater incentive than other undertakings to challenge doubtful patents and so avoid the obligations and restrictions incurred under the licence. In *AOIP* v. *Beyrard*,[60] the Commission added that the licensee might be the person best placed to challenge the patents on the basis of information given to it by the licensor. In several individual decisions, the Commission condemned restrictions on challenging not only patents, but also trade marks.[61]

The approach adopted by the Commission in its older decisions was, however, static. Once the licence is granted, the market would be more competitive if invalid patents could be challenged. What the analysis overlooks, however, is the effect of such a rule on the holder's willingness to grant licences.[62] Patent litigation is notoriously lengthy and expensive and licensors, especially small firms, may not be able to survive it. A patentee may well not be able to obtain firm advice on the validity of its patent. It may not know of a prior publication of the innovation, and the test of obviousness or its converse, an inventive step, is inherently subjective.

If a licensee is likely to challenge a patent,[63] whether or not the licensor

---

[60] *AOIP/Beyrard*, [1976] OJ L6/8, [1976] 1 CMLR D14, CMR 9801. *Davidson* is cited at n. 16 above.

[61] I discussed the older decisions rather more fully in **Valentine Korah**, *Patent Licensing and EEC Competition Rules—Regulation 2349/84* (Oxford: ESC Publishing, 1985), 61–5. The book is out of print, but available in some libraries.

[62] See 1.3.1 above for the Commission's greater willingness recently to analyse transactions *ex ante*.

[63] A patent or the secrecy of the know-how is not likely to be challenged before the licensee has invested in production facilities. So, often, a licensee has an incentive not to challenge in order to exclude third parties. Nevertheless, with the passage of time, it may want to use rival technology, or just refuse to pay royalties. This final option may be avoided by providing that, in the event of challenge, the licence may be determined.

believes it to be valid, the holder may well hesitate to grant a licence. Indeed, since the decision in *Davidson Rubber*, many firms have granted few licences for the common market countries to firms that they did not control, whether through a shareholding or some commercial relationship. The inability to restrain challenge must have reduced the incentive both to licence existing technology and to invest in innovation. A holder may prefer to protect the territory for which it can produce than license a third party for a marginal area for which it lacks production capacity. Usually more is to be made from making and selling than from licensing.

The Court upheld the Commission's *per se* condemnation of a no-challenge clause relating to a trade mark that had not been licensed in *Windsurfing*,[64] although Advocate General Lenz had pointed out that the clause had no significant effect on the competitive position of licensees which had developed their own marks, and so had no significant effects on competition. The judgment may be distinguished on the ground that usually the mark not to be challenged is one being licensed. Moreover, it is quite possible that the mark in *Windsurfing* had become descriptive of the sport, so was particularly likely to be invalid. The strong-minded may also argue that the judgment was given by a chamber of only three judges and could well be reversed by the full Court. Indeed, it may already have been in *Bayer and Hennecke*,[65] considered shortly below, and is inconsistent with the Commission's subsequent decision in *Moosehead/Whitbread*.[66]

In *Windsurfing*,[67] the Court followed the Advocate General in relation to the restriction on challenging the validity of the patent. Herr Lenz pointed out that some other licensees had not been restrained, but had not used their liberty to challenge the validity of the patent. Moreover, after the abrogation of the clause, the two licensees who had been restrained had not challenged it either. Herr Lenz observed that a licensee may well not want to challenge a clause that would open its market up to competitors, so the restriction on its freedom to challenge might not be significant. Nevertheless, he thought the Commission's condemnation could not be attacked.

More recently the Court has been more flexible in its approach to no-challenge clauses in patent licences. In *Bayer and Hennecke*, both litigation and opposition proceedings in the German patent office were compromised on the grant of reciprocal non-exclusive licences under the

---

[64] (193/83), [1986] ECR 611, [1986] 3 CMLR 489, CMR 14,271, comment **James Venit**, n. 7 above, 558. That judgment, however, was delivered by a Chamber of only three judges, when Judge Joliet was recovering from surgery and Judge O'Higgins, as the most junior member of the chamber, stood down to leave an uneven number.

[65] *Bayer and Hennecke* v. *Süllhöfer* (65/86), [1988] ECR 5249, [1990] 4 CMLR 182, [1990] 1 CEC 220, a decision of a full Court of nine. See comment by **Valentine Korah**, (1988) 12 *EIPR* 381, 384.   [66] N. 59 above.   [67] N. 29 above.

German patent and model containing a no-challenge clause relating to the validity of the corresponding industrial property rights of Bayer and Hennecke in other Member States. The Community Court was asked to give a preliminary ruling on the validity of the no-challenge clause. The Commission argued that, although it was still hostile to restrictions on challenging intellectual property rights, there was a public interest in compromising litigation, and that the no-challenge clause was ancillary to the compromise, not *vice versa*. In such circumstances the clause should fall outside article 85(1).

This the Court rejected. It ignored the precedents on ancillary restraints and said that article 85 draws no distinction between agreements made to compromise litigation and those made for other purposes. Nevertheless, both Advocate General and Court pointed out that, before finding that a no challenge clause infringes article 85(1), the national judge should look to its legal and economic context. The Community Court is not required to apply its ruling under article 177: that task is for the national court that requested a preliminary ruling.

The judgment is important for deciding that such clauses are not necessarily illegal, although the examples the Court gave of situations where a restraint on challenge would not infringe article 85(1) were not helpful. At paragraph 17, the Court suggested that such a clause would not restrict competition if the licence were royalty free, since the licensee would not be under a competitive disadvantage. This misses the point that the patent may exclude others, who have more incentive to challenge, but may be less able to do so. Moreover there was consideration for the licence in the grant of a reciprocal licence.

The Court seems to have been more concerned with fair than free competition. The particular licence seems to have been granted to a competitor, in which case reciprocal licences of an invalid patent together with a no-challenge clause relating to patents elsewhere might be a cosy way of excluding competitors. Most disputes that lead to licences are between competitors, so a difficult decision was needed on whether the possibility of restricting competition by allocating markets between competitors outweighed the cost savings from resolving the dispute, enabling the parties to carry on producing, or preparing to produce, without concern about infringing each other's intellectual property rights. Given that it is at the time the agreement is being negotiated that certainty is required, it is thought that the Commission was right to argue that obligations not to challenge are ancillary to the compromise and should automatically be cleared.

The *Bayer and Hennecke* case makes it extremely difficult to resolve disputes over the validity of intellectual property rights. A compromise, under which the holder of the possibly valid right provides consideration to the person opposing validity, often includes a restriction on the latter challenging the right. It has become more difficult to argue that such a restraint is justified by the saving in the costs of litigation and the reduction of uncertainty.

Where such a compromise is made between firms which could have competed without any licence, such an agreement could be a sham, concealing a horizontal arrangement that permits a limited amount of territorial protection. Where, however, it is not a sham, such an agreement may save a great waste of resources. It is, however, not easy to tell when an agreement is a sham.

The Court states that the agreement must be examined in its economic and legal context before it is found to infringe article 85(1). So, if it can be shown that the dispute had merit on both sides, it would be arguable that the no-challenge clause does not infringe article 85(1). Nevertheless, the cautious may decide to notify such settlements to the Commission as a matter of course, and hope that the Commission will decide not to oppose the group exemption.

It is thought that a recital in a licence agreement acknowledging that the know-how is valuable and secret at the time of the licence and that the patent licensed is valid does not amount to a 'prohibition' on contesting its secrecy and may be helpful in altering the onus of proof in a national court when it is decided not to notify under the opposition procedure.[68]

7.3.7 Royalties After the Know-how Enters the Public Domain or the Licensed Patents Expire, on Inappropriate Products, and the Unilateral Prolongation of the Agreement (Articles 2(1)(7) and 8(3))

Know-how is precarious in that, as it becomes known to those in the industry, it loses its value because there is no exclusive right to enforce. The holder may seek to pass on the risk of premature disclosure, or part of it, to his licensee(s). The licensee may also find it easier to finance the initial payments needed to establish production and develop a market if it can pay lower royalties or down payments in the early years in return for payments that continue.

---

[68] In *Computerland* [1987] OJ L114/2, [1988] 4 CMLR 592, CMR 12,165, at para. 5, the Commission noted that there was an acknowledgement in a franchise agreement that a mark is valid. It does not refer to this in its legal assessment, from which one may infer that such an acknowledgement does not infringe art. 85(1). In *Rich Products/Jus-Rol*, n. 18 above at para. 11, the agreement stated that know-how disclosed to the licensee 'remains the property of the licensor' and this provision was not analysed under art. 85, so the latter was, presumably, not considered relevant.

In *Moosehead /Whitbread*, n. 59 above, Whitbread acknowledged Moosehead's title to the marks and the validity of the registration (point 8.3), but this was treated in the legal assessment as amounting to a no-challenge clause (point 15.4).

Article 2(1)(7) permits an obligation to continue to pay royalties, and/or contractual damages, if the know-how has become public otherwise than through the action of the licensor. It whitelists:

> (7) an obligation on the licensee to continue paying the royalties:
>
> (a) until the end of the agreement in the amounts, for the periods and according to the methods freely determined by the parties, in the event of the know-how becoming publicly known other than by action of the licensor, without prejudice to the payment of any additional damages in the event of the know-how becoming publicly known by the action of the licensee in breach of the agreement;
>
> (b) over a period going beyond the duration of the licensed patents, in order to facilitate payment.

This follows the same provision in the know-how regulation, which represented a change in thinking by the Commission. It no longer intervenes to help a licensee who may find it difficult to compete with others who are under no obligation to pay royalties.[69] Even in 1987, when the draft of the know-how regulation was published, an obligation to pay was permitted for only three years after the know-how became public.[70] The Commission now acknowledges that royalties are a matter to be determined freely by the parties. A licensee who will continue to be liable for royalties after disclosure of the know-how may be more careful not to disclose it, and might not have been able and willing to pay in advance or in the early years as much as can be recovered as royalties towards the end.

The Commission explains its thinking in recital 21 of regulation 240/96:

> (21) The list of clauses which do not prevent exemption also includes an obligation on the licensee to keep paying royalties until the end of the agreement independently of whether or not the licensed know-how has entered into the public domain through the action of third parties or of the licensee himself (article 2(1)(7)). Moreover, the parties must be free, in order to facilitate payment, to spread the royalty payments for the use of the licensed technology over a period extending beyond the duration of the licensed patents, in particular by setting lower royalty rates. As a rule, parties do not need to be protected against the foreseeable financial consequences of an agreement freely entered into, and they should therefore be free to choose the appropriate means of financing the technology transfer and sharing between them the risks of such use. However, the setting of rates of royalty so as to achieve one of the restrictions listed in Article 3 renders the agreement ineligible for the block exemption.

---

[69] See *AIOP* v. *Beyrard*, cited at n. 60 above, criticized by **Valentine Korah**, (1976) 3 *EL Rev.* 185; where the Commission condemned a duty to pay royalties once the last patent existing at the time of the licence had expired. Contrast *Rich Products/Jus-Rol*, cited at n. 18 above.   [70] [1987] OJ C214/2, art. 2(1)(9).

Since the introductory words of article 2 refer to article 1, and article 1 applies to mixed know-how and patent licensing agreements, it is clear that this recital applies to pure patent licensing agreements without any unprotected know-how. This is a minor but welcome change.

The Commission used to argue that an unduly long period of paying royalties for a patent licence made it more difficult for the licensee to compete after the term. Now, however, it accepts that if the licence has been negotiated with a sensible licensee, it will insist on paying less if it agrees to pay for longer. If a smaller royalty is payable on its initial production the licensee's initial risk may be reduced. It may be able to pay a little extra later, when it has already started production and sale and is enjoying some lead time over its competitors. This freedom to accept liability for continuing royalties, unless the know-how has become public through the fault of the licensor, is a considerable improvement on the Commission's earlier thinking. It is another example of the Commission beginning to think *ex ante* to the time when the transaction is agreed, rather than when the Commission is looking at the agreement.[71]

I doubt whether royalties may be demanded for an indefinite period. Recital 21 seems to apply to a fixed period of time that is reasonable in all the circumstances. The Commission argued in its written submissions to the Community Court in *Ottung* v. *Klee*[72] that the parties should be free to agree royalties as they wish:

> As for the payment of royalties, the Commission considers in particular that their payment after the patent has expired does not fall within Article 85(1) when that form of payment is adopted in order to facilitate payment or constitutes a means of calculating the royalties. In the Commission's view, the same is true when a licensing contract remains in force after the patent or the intellectual property right initially granted in the licence has expired and the parties had agreed that the licensee would continue to pay royalties after the expiry of the patent for as long as the contract remained in force. Referring to the judgment [in *Windsurfing*], the Commission maintains that in such a case the parties are free to calculate the royalty as they wish.
>
> Consequently, the Commission considers that Article 85(1) must be interpreted as meaning that a clause in a licensing agreement whereby the parties agree that the licensee will pay to the grantor of the licence a royalty calculated on the basis of turnover as payment for the right to exploit the invention after the expiry of the patent granted by the licensing agreement and for as long as the agreement remains in force is not a restriction of competition within the meaning of that Article.

[71] See 1.3.1 above.
[72] (380/27), [1989] ECR 1177, [1990] 4 CMLR 915, [1990] 2 CEC 674, at 1183 of the ECR.

The Court ruled that:

> 2. A contractual obligation under which the grantee of a licence for a patented invention is required to pay royalty for an indeterminate period, and thus after the expiry of the patent, does not in itself constitute a restriction of competition within the meaning of Article 85(1) of the Treaty, where the agreement was entered into after the patent application was submitted and immediately before the grant of the patent.
>
> 3. A clause contained in a licensing agreement prohibiting the manufacture and marketing of the products in question after the expiry of the patent and the termination of the agreement comes within the prohibition laid down in Article 85(1) only if it emerges from the economic and legal context in which the agreement was concluded that it is liable to appreciably affect trade between Member States.[73]

In its judgment, the Court made the qualifications suggested by the Advocate General, that the obligations relating to the period after expiry of the patents must be in return for the licence given before, but it is unlikely that the licensee would agree to them unless that be the case. It is hoped that the licensee is not required to establish this.

In the know-how regulation, article 2(1)(7) was limited by article 3(5), which blacklisted licences where the licensee was required to pay royalties on products not even partially produced by means of the licensed technology. This provision has been deleted in the technology transfer regulation, as has article 7(7) which empowered the Commission to withdraw the benefit of the group exemption when the royalties continued too long after the know-how became public. It seems to me that the Commission has changed its thinking dramatically, and that such an obligation may not now even be subject to the opposition procedure.

Article 8(3) has also reversed the position on the automatic prolongation of the licence by the addition of new improvements by the licensor, provided that the licensee has a right to refuse such improvements or each party has the right to terminate the agreement at the expiry of the initial term and at least every three years thereafter. Under the know-how regulation, article 3(10) blacklisted licences when the initial period could be prolonged by the addition of new improvements, unless the licensee had the right to refuse the improvements or each party had the right to terminate. Now it is clear from article 8(3) that the regulation applies to agreements providing for automatic prolongation, subject to the provisos.

> 3. This Regulation shall furthermore apply to pure patent or know-how licensing agreements or to mixed agreements whose initial duration is

---

[73] In setting out its reasons, the Court added at para. 12 that whether such a clause restricts competition would also depend on the legal and economic context of the agreement.

automatically prolonged by the inclusion of any new improvements, whether patented or not, communicated by the licensor, provided that the licensee has the right to refuse such improvements or each party has the right to terminate the agreement at the expiry of the initial term of an agreement and at least every three years thereafter.

7.3.8 Field-of-use, Customer Restrictions, and Maximum Quantities (Articles 2(1)(8), 3(4), and 4(2))

When the Commission believed that an agreement was horizontal once licensor and licensee were producing the licensed products,[74] it was hardly surprising that it considered the allocation of customers to be anti-competitive. One way of dividing markets is to arrange that a licensee shall serve only one kind of customer, leaving others to the licensor or other licensees. Where, however, the agreement is vertical, in the sense that the licensee could not have competed but for the licence, it is unlikely that allocation of different products or customers is intended to restrict production and raise prices. It may be the only, or best, way of ensuring that each kind of customer or product is supplied by someone sufficiently protected from competition in the same technology to induce it to invest in tooling up and making a market.

Already by the time the patent regulation was adopted, under pressure from industry, the Commission had begun to change its mind about field-of-use provisions. Quantity limitations and the allocation of customers were blacklisted, but the white list included a field-of-use clause, drafted in terms that were difficult to distinguish from the blacklisted clause.

7.3.8.1 *Permissible Field-of-use Restriction (Article 2(1)(8))*

The field-of-use restriction whitelisted by article 2(1)(8) of the know-how and technology transfer regulations extends further than article 2(1)(3) of the patent regulation in that the licensee may be limited not only to a technical field of application covered by the licensed technology, whatever that may mean, but also to particular products. Article 2(1)(8) permits:

> (8) an obligation on the licensee to restrict his exploitation of the licensed technology to one or more technical fields of application covered by the licensed technology or to one or more product markets;

Recital 22 explains:

> (22) An obligation on the licensee to restrict his exploitation of the licensed technology to one or more technical fields of application ('fields of use')

---

[74] See 1.4.2 above.

or to one or more product markets is not caught by Article 85(1) either, since the licensor is entitled to transfer the technology only for a limited purpose (Article 2(1)(8)).

In *Delta Chemie/DDD*, [75] the Commission stated that a restriction on the licensee using the know-how so long as it remains secret only to make the licensed products does not restrict competition, as it is the corollary of the licensor's right to dispose freely of its know-how. Otherwise the holder would be deprived of a greater or lesser part of the income to be derived from its know-how. This seems to be another example of the Commission beginning to accept the concept of a limited licence.[76] The licensee is given the technology for use only in producing certain products. It is also an example of the Commission accepting, and extending to know-how, the view of the Court in *Coditel II*,[77] *Warner* v. *Christiansen*,[78] and *Erauw-Jacquéry*[79] that the holder of copyright or plant breeders' rights must be able to obtain fair remuneration if it is to be induced to invest.

The Commission might have added in the recitals to the know-how and technology licensing regulations that field-of-use restrictions enable the innovator to ensure that each market is properly exploited. It has always seemed anomalous to me that the Commission has recognized the need in vertical transactions for exclusive territories, but not for other ways of dividing markets that might encourage a licensee or distributor to provide services for the benefit of the brand as a whole without fear of others taking a free ride on his investment. Customer restrictions are blacklisted only if the parties were competing manufacturers, but quotas are blacklisted even if they were not. Like field-of-use restrictions they are less likely to isolate national markets than exclusive territories. The Commission still shows signs of believing that technology licences are horizontal agreements.

Despite the relaxations permitting a licensee to be licensed for only specified fields of use or products, and providing for a second source, the Commission is not happy about field-of-use restrictions in horizontal licences. In *Decca Navigator Systems*,[80] Racal-Decca, a firm found to be dominant in supplying receivers to pick up the signals it transmitted to guide mariners, licensed AP to provide receivers for the signals transmitted

[75] N. 10 above, para 31.

[76] See 3.2–3.2.5 above. The Commission does not go all the way, since customer restrictions are blacklisted when not justified as field-of-use restrictions.

[77] *Coditel II—Coditel SA and Others* v. *Ciné-Vog Films and Others SA* (262/81), [1982] ECR 3381, [1983] 1 CMLR 49, CMR 8862, 1.4.1, text to n. 54, above.

[78] *Warner Bros. Inc. and Metronome Video ApS* v. *Christiansen* (158/86), [1988] ECR 2605, [1990] 3 CMLR 684, [1990] 1 CEC 33.

[79] *Erauw-Jacquéry (Louis) SPRL* v. *Société La Hesbignonne* (27/87), [1988] ECR 1919, [1988] 4 CMLR 576, [1989] 2 CEC 637, also discussed at 1.4.1 above.

[80] *Decca Navigator System*, n. 58 above.

as an aid to navigators, after AP had already started to sell a DNS-compatible receiver. The licence provided that AP should supply Decca receivers only for pleasure boats, with restrictions on AP's customers using or selling the receiver for use in other kinds of boat.

Since AP was already able to make a receiver, it was a competitor of Racal Decca, and the Commission's antagonism towards a horizontal agreement to limit the entry of a new competitor to the pleasure-boat market may have been justified.[81] AP was also restrained from making the more sophisticated receivers needed for professional navigation.

7.3.8.2 *The Black List, Customer Allocation (Article 3(4))*

As in the earlier regulations, article 3(4) prevents the application of the regulation if customers are allocated as between licensor and licensee. Recital 23 explains:

> (23) Clauses whereby the parties allocate customers within the same technological field of use or the same product market, either by an actual prohibition on supplying certain classes of customer or through an obligation with an equivalent effect, would also render the agreement ineligible for the block exemption where the parties are competitors for the contract products (Article 3(4)). Such restrictions between undertakings which are not competitors remain subject to the opposition procedure.

This is implemented by article 3(4). The regulation does not apply where:

> (4) the parties were already competing manufacturers before the grant of the licence and one of them is restricted, within the same technical field of use or within the same product market, as to the customers he may serve, in particular by being prohibited from supplying certain classes of user, employing certain forms of distribution or, with the aim of sharing customers, using certain types of packaging for the products, save as provided in Article 1(1)(7) and Article 2(1)(13);

A major change in the technology transfer regulation is that the black list applies only where the parties were competing manufacturers before the grant of a licence, although recital 23 says that where they are not, the agreement is not exempted unless notified under the opposition procedure and not opposed by the Commission.

'Competing manufacturers' is defined in article 10(17):

---

[81] The Commission stated that only from Apr. 1987 did the General Lighthouse Authority pay the cost of transmitting the signals. So, until then, Racal Decca did need to use some device to prevent the makers of receivers taking a free ride on its signals. In failing to address this problem the decision is less than satisfactory.

(17) 'competing manufacturers' or manufacturers of 'competing products' means manufacturers who sell products which, in view of their characteristics, price and intended use, are considered by users to be interchangeable or substitutable for the licensed products.

Competing is defined only in terms of substitutes on the demand side of the market. Ease of entry, or potential competition, is irrelevant to the application of article 3(4), although it would be relevant to the Commission's appraisal if the agreement be notified under regulation 17 or under the opposition procedure. It is only where the firms are already manufacturing substitutes at the time of the licence that the black list will apply.

Since 'technical field of application' is not a term of art and there is as yet no case law, in some cases it will continue to be impossible to advise firmly whether particular clauses fall into the white or black list. Field-of-use restrictions will always allocate customers or products. The field of use should be specified in the original licence, and it may help to call it such: a licence might be confined to 'veterinary use' rather than 'for the supply to vets'. The problem will continue to arise since customer restrictions will prevent the regulation applying, unless the agreement be notified even if the parties were not 'competing manufacturers' at the time of the licence.

Different product markets is also a concept that may give rise to difficulty. In article 10(17) of the technology transfer regulation, the notice on minor agreements,[82] and article 3(a) and (b) of the regulation exempting exclusive distribution, product markets are defined in terms of substitutes on the demand side—'manufacturers of identical goods or of goods which are considered by users as equivalent in view of their characteristics, price and intended use'.[83] Both concepts may give rise to difficulty in practice, but the whitelisted clause is broader than under the patent licensing regulation, in that it applies to separate product markets as well as to fields of use and the black list applies only between competing manufacturers. This liberality is welcome.

As under the earlier regulations, the parties may not indirectly allocate customers by requirements as to packaging. Otherwise one licensee might be permitted to sell in bulk packages and another in individual ones; one in packages labelled suitably for the veterinary trade and another as required for the treatment of human beings. Such restrictions would come within article 2(1)(8) only where they are justified by a field-of-use or product restriction, since it is clear that the whitelisted provision prevails. Article 3(4) circumscribes it where the parties were competing manufacturers.

---

[82] [1984] OJ C231/2, as amended [1994] OJ C368/20.

[83] An economist would consider such products as substitutes only if they were in the same geographic market. The notice provides for this to a limited extent, but reg. 1983/83 does not.

Unless the products can be sufficiently differentiated to come within article 2(1)(8), one cannot divide markets by requiring licensees for the high-priced market to adopt higher quality standards than those for the low-priced market without subjecting the licence to the opposition procedure. As explained at 7.3.5 above, article 2(1)(5) permits the licensor to require quality standards not needed for a technically satisfactory exploitation of the licensed technology only if the standards are also respected by the licensor and other licensees. This provision is tightly circumscribed by article 4(2)(a) unless the licensees want at the date of their agreement to be bound to follow the quality specifications.

Where the licensor wants only some customers to obtain the protected products, it cannot restrain its licensees from supplying other customers without the agreement being caught by article 3(4) if the parties were competing manufacturers or being subjected to the opposition procedure, if not. It seems, however, that it can grant a licence to selected customers to make or have the products made solely for their own use under article 1(1)(8) and the selected customers can have them made by taking advantage of the sub-contracting notice. The holder of the technology would not then be able to select the sub-contractor. The regulation is drawn in terms that enable those who are well advised to do things indirectly and qualify under the group exemption.

7.3.9 Minimum Royalties and the Obligation Not To Compete (Articles 2(1)(9), 3(2), 2(1)(17) and (18), and 7(4))

Where an exclusive licence is granted, and most licences exempted by this regulation will confer an exclusive territory on the licensee with associated restrictions on other licensees exploiting the territory, the licensor has an incentive to ensure that its technology is not neglected. It may encourage the exploitation of its technology through a 'best endeavours' clause, a payment or payments in advance as in *Boussois/Interpane*,[84] or by stipulating for a minimum royalty or some measure of minimum use of the technology.

On the other hand, by discouraging the exploitation or development of rival technology, incentives to use the licensed technology might discourage the licensee from using and developing rival technology. According to the US guidelines on intellectual property licensing, a rule of reason is being applied in that jurisdiction to balance these considerations. The guidelines state:

---

[84] Cited at n. 3 above, discussed at 3.1.3 above, where the Commission cleared an obligation to pay lump sums at various stages before production commenced.

5.4 Exclusive dealing

In the intellectual property context, exclusive dealing occurs when a license prevents the licensee from licensing, selling, distributing, or using competing technologies. Exclusive dealing arrangements are evaluated under the rule of reason. . . . In determining whether an exclusive dealing arrangement is likely to reduce competition in a relevant market, the Agencies will take into account the extent to which the arrangement (1) promotes the exploitation and development of the licensor's technology and (2) anticompetitively forecloses the exploitation and development of, or otherwise constrains competition among, competing technologies.

The likelihood that exclusive dealing may have anticompetitive effects is related, inter alia, to the degree of foreclosure in the relevant market, the duration of the exclusive dealing arrangement, and other characteristics of the input and output markets, such as concentration, difficulty of entry, and the responsiveness of supply and demand to changes in price in the relevant markets. . . . If the Agencies determine that a particular exclusive dealing arrangement may have an anticompetitive effect, they will evaluate the extent to which the restraint encourages licensees to develop and market the licensed technology (or specialised applications of the technology), increases licensors' incentives to develop or refine the licensed technology, or otherwise increases competition and enhances output in a relevant market.

7.3.9.1 The White List, Minimum Royalty or Quantity (Article 2(1)(9))

Article 2(1)(9), like the same provision in the earlier regulations, permits:

> (9) an obligation on the licensee to pay a minimum royalty or to produce a minimum quantity of the licensed product or to carry out a minimum number of operations exploiting the licensed technology;

Where an exclusive licence is given, the licensor may need to ensure that his technology is not sterilized in the territory. In my view, the obligation to pay lump sums early on has the same objective and should be treated similarly. Such an obligation was cleared in *Boussois/Interpane*,[85] before the adoption of the know-how regulation.

Where the parties were competitors before the grant of the licence, however, such provisions may restrict the licensee from using competing technologies, and this is one of the grounds on which the Commission may withdraw the benefit of the exemption. See 7.3.9.5 below.

7.3.9.2 The Black List, Not to Compete (Article 3(2))

Article 3(2), like provisions in the earlier regulations, prevents the exemption from applying where one party is restricted from competing

---

[85] N. 3 above.

with the other or with other undertakings in respect of r & d, production, or use of competing products, and their distribution: the exemption does not apply where:

> (2) one party is restricted from competing within the common market with the other party, with undertakings connected with the other party or with other undertakings in respect of research and development, production, use or distribution of competing products without prejudice to the provisions of article 2(1)(17) and (18);

This provision is more qualified than the equivalent provision in the earlier regulations. It makes it clear that it is without prejudice to article 2(1)(17) and (18), discussed at 7.3.9.3 and 7.3.9.4 below. I think that article 3(2) is also subject to article 2(1)(9). Since the Commission can withdraw the benefit of the exemption under article 7(4) where the parties are competitors and agree on minimum royalties etc., there seems to be little risk in permitting the whitelisted provision to prevail.

A restriction on competing outside the common market will infringe article 85(1) only if it has perceptible effects within it and is not blacklisted. According to recital 7, such a restriction does not prevent the regulation applying to restrictions relating to the common market, presumably, even if the restriction does restrict competition within the common market.

7.3.9.3 Licensee's Best Endeavours (Article 2(1)(17))

As in the earlier regulations, the Commission seems to have been unable to make up its mind. It has whitelisted the licensee's obligations to use its best endeavours to exploit the licensed technology. Article 2(1)(17) whitelists:

> (17) an obligation on the licensee to use his best endeavours to manufacture and market the licensed product;

For the licensee to compete in exploiting alternative technology contrary to article 3(2) would infringe most best-endeavours clauses. The two kinds of provision have identical effects, yet the former is blacklisted and the latter whitelisted! If the licensee wanted to use rival technology, it would probably notify the Commission, or threaten to do so, and renegotiate the licence to avoid the exemption being withdrawn.

When scrutinizing technology licences, business lawyers merely remove the blacklisted clauses, and substitute some of the provisions whitelisted by article 2(1)(9), (17), and (18). Article 3(2) may remain effective in relation to developing rival technology, although even to do that might be considered contrary to a whitelisted promise by the licensee to use its best endeavors to exploit the licensed technology. Where the parties were already competitors when the licence was granted, that might be a reason for the Commission to withdraw the benefit of the exemption under article 7(4). It is hoped that it would not do so under the general words of article 7 when they become competitors as a result of the licence.

7.3.9.4 *The Right to Terminate Exclusivity and Cease Feeding Further Know-how (Article 2(1)(18))*

The restriction on competing with the other party, blacklisted by article 3(2), was subject to two other provisos in the know-how regulation. It was not clear whether these were restrictions of competition subject to the opposition procedure. It is now clear that they are not. Article 2(1) whitelists:

> (18) a reservation by the licensor of the right to terminate the exclusivity granted to the licensee and to stop licensing improvements to him when the licensee enters into competition within the common market with the licensor, with undertakings connected with the licensor or with other undertakings in respect of research and development, production, use or distribution of competing products, and to require the licensee to prove that the licensed know-how is not being used for the production of products and the provision of services other than those licensed.

These provisos considerably reduce the problems resulting from black-listing a restriction on competing. Know-how is vulnerable to disclosure and misuse. It may be important to ensure that it cannot be used with other technology and fed back to a competitor under a feedback clause or used by a licensee who fails to pay royalties. If desired, the rights under these provisos should be written into the original licence.

Fortunately, the Commission recognized in the know-how regulation that know-how that is not protected by patents is more vulnerable than technology protected by exclusive property rights. The licensor may terminate the exclusivity in the event of the licensee competing in any of the ways mentioned, and may require the licensee to prove that the licensed know-how was not used for the production of goods and services other than those licensed.

Nevertheless, the licence will continue, preventing the licensor from granting a fully exclusive licence should the first licensee start to use or develop rival technology. Licensing contracts should make it clear whether there is a duty to carry on feeding know-how in such circumstances. There is no longer any question of the need to notify such a provision under the opposition procedure.

These whitelisted provisions go further than article 2(1)(9) of the technology transfer regulation, in that they apply not only to manufacture and marketing, but also to development of rival technology, although a minimum royalty etc. might well discourage such development.

The know-how and technology transfer regulations enable the licensor to do more to protect his technology than did the patent regulation. In part, this is due to the Commission's acceptance that know-how is valuable, but more vulnerable.

In *Delta Chemie/DDD*,[86] the Commission cleared the licensee's obligation not to produce or distribute goods similar to those made under licence without consent, which was not to be withheld if there was sufficient certainty that the licensee would respect its contractual obligations. That agreement, however, permitted the licensee to make competing products as a subcontractor for clients who sold them through other means. The Commission stated that the clause enabled the licensor to ensure that the licensee did not use the know-how to make products other than those envisaged. Moreover the clause would encourage it to use its best endeavours to exploit the know-how and sell the licensed products.

As described at 7.3.1 above, a duty on the licensee not to divulge the licensor's technology is clearly whitelisted by article 2(1)(1), and this may inhibit the licensee from taking a second licence under rival technology, where it would be required to feed back improvements to the second licensor which might divulge the technology obtained under the first licence. Under article 2(1)(4), discussed at 7.3.4 above, a licensee may be required to feed or grant back improvements to the first licensor only if it is free to use its own improvements, and to license them to others, if licensing does not involve disclosing the first licensor's technology. The licensor's consent to such a licence may be refused only on grounds that its technology may be disclosed.

7.3.9.5 Commission's Power to Withdraw the Exemption (Article 7(4))

A best-endeavours clause and an obligation to pay a minimum royalty etc. are ways in which a cartel could be formed behind the facade of a licence of trivial technology. The restrictions relate not to the technology licensed and products made thereby, but to rival technology. If the parties could have competed without the licence, so that even the US authorities would treat the agreement as horizontal, a best-endeavours clause may restrain the licensee from using rival technology.

Article 7(4), therefore empowers the Commission to withdraw the benefit of the group exemption where:

> (4) the parties were competing manufacturers at the date of the grant of the licence and obligations on the licensee to produce a minimum quantity or to use his best endeavours as referred to in Article 2(1), (9) and (17)

---

[86] N. 10 above.

respectively have the effect of preventing the licensee from using competing technologies.

This seems to me a sensible provision. Only where the parties were already competing manufacturers[87] will article 7(4) apply. It is true that under the introductory words the exemption can be withdrawn whenever the Commission thinks an agreement does not merit exemption, but the specific examples given must be intended to warn industry of the Commission's concerns. Since the Commission enjoys a broad discretion when withdrawing the exemption, it can consider whether entry barriers are high, and the other matters specified in the US guidelines, quoted at 7.3.9 above.

7.3.10 Most-favoured Licensee Terms (Article 2(1)(10))

Article 2(1)(10) is identical to article 2(1)(11) of the patent regulation. It permits:

(10) an obligation on the licensor to grant the licensee any more favourable terms that the licensor may grant to another undertaking after the agreement is entered into;

This item seems to be based on the ideas of fair competition that colour thinking in the EC. Many small firms believe that discrimination is unfair and should be restrained. Nevertheless, where markets are tightly oligopolistic, the only kind of price competition commercially possible is secret discounts that cannot easily and quickly be matched by the other suppliers. In such a market most-favoured licensee provisions may restrict competition. A firm is less likely to give a discount to one customer if it has to be matched by others. There are three ways in which 'most-favoured customer' clauses can have horizontal effects: they may constitute facilitating practices, raise rivals' costs, and dampen competition.[88] Most of the analysis applies equally to licences.

---

[87] Defined in art. 10(17), analysed at 7.3.8.2 above.

[88] See **Jonathan B. Baker**, 'Vertical Restraints with Horizontal Consequences: Competitive Effects of "Most-Favored-Customer" Clauses', Speech before Business Development Associates, Inc, Antitrust Conference 1996. As a facilitating device, Baker argues that oligopolistic co-ordination works better when the incentive to cheat is reduced. A firm adopting a most-favoured customer policy might find it hard to cheat on an obligation to give price concessions to others, and therefore have less incentive to give the initial discount. By announcing such a policy, it might also become the price leader. It may suffice to announce such a policy only to one or a few major customers.

A most-favoured customer clause may raise the costs of the customer's rivals, who will find it more difficult to negotiate better deals. The issue has arisen in the US in relation to health care insurance plans. Such a clause may also dampen competition between the suppliers, especially if adopted by many of them. Since the supplier is deterred from granting discounts

In *Kabelmetal*,[89] the Commission cleared an obligation not to grant more favourable terms to other licensees, but observed that in other circumstances such an obligation might inhibit licensing on the only terms available. The Commission's qualification rests on a static analysis. In that case the goods were custom-made and did not pass through dealers. In the case of products that are traded, an exclusive licensee may need more protection from dealers buying on more favourable terms from a licensee liable for lower royalties, as such goods can pass freely through the common market. Without promising treatment as favourable as that given to any other licensee, it might not be possible for the holder of technology to persuade a licensee to invest. The Commission is now prepared to look *ex ante* to the time when a transaction is being negotiated,[90] so it is hoped that the old qualification in *Kabelmetal* can be forgotten. In oligopolistic markets, where Mr Baker's concerns may be relevant, the Commission can withdraw the benefit of the exemption if it receives complaints from rivals whose costs are increased or purchasers from the licensees.

Where the first licensee is expected to develop the system and pass back improvements to be disseminated to later licensees, it may need to be assured of more favourable terms than are given to subsequent licensees. The latter are likely to have advantages the first licensee lacked. They may be able to send potential customers to view the original plant and its produce, as well as set up a more efficient plant as a result of the first licensee's improvements. I hope, therefore, that the promise of a discounted royalty or smaller payment in advance might be treated as the counterpart of the first licensee's services, from which the licensor and subsequent licensees will benefit. Consequently, a promise that the royalties paid by the first licensee will be lower than those payable by other licensees may not go beyond what is whitelisted by article 2(1)(10). There is, however, no authority for this and it might be safer for the licensor to make a separate payment for the services performed by the first licensee, and promise only not to give more favourable terms to subsequent licensees, without offering them to the first.

7.3.11 An Obligation to Mark (Articles 2(1)(11) and 1(1)(7))

The eleventh item on the whitelist is:

> (11) an obligation on the licensee to mark the licensed product with an indication of the licensor's name or of the licensed patent;

by the need to spread the discount, it indicates that it will not compete aggressively. Rivals may respond similarly. This is most likely to dampen competition when there are few firms and higher prices would not lead to new entry.

No one favours making most-favoured customer clauses illegal *per se*, but Mr Baker suggests that the efficiencies should be balanced against the possible harms to competition.

[89] N. 17 above.   [90] See 1.3.1 above.

The corresponding provision in article 2(1)(11) of the know-how regulation did not permit an obligation to mark the licensed product with an indication of the licensed patent, but the non-exhaustive nature of the white list may have embraced such a provision. An obligation to use the licensor's mark or get up is exempted by article 1(1)(7), analysed at 6.6.3 above. A restriction on the licensee identifying himself as the manufacturer, which is not exempted by article 1(1)(7), is not blacklisted but would probably need substantial justification as an unlisted restriction if notified under the opposition procedure.

In *Burroughs/Delplanque*,[91] the Commission cleared an obligation to mark on the ground that it facilitated checks on quantity and quality. 'The licensed products' are defined in article 10(8) as:

> goods or services the production or provision of which requires the use of the licensed technology;

'Licensed technology' is defined by article 10(7) to mean

> the initial manufacturing know-how or the necessary product and process patents, or both, existing at the time the first licensing agreement is concluded, and improvements subsequently made to the know-how or patents, irrespective of whether and to what extent they are exploited by the parties or by other licensees;

Where some of the technology ceases to be sufficiently secret to qualify for the exemption, care should be taken to ensure that the products to be marked are covered by patents or by know-how that remains secret and substantial.

According to recital 10 of the patent regulation, the licensee had to be permitted to indicate that he had manufactured the products. There is no equivalent recital in the later regulations, so the question arises whether the recitals in the patent regulation apply to the provisions that have followed its wording.

7.3.12 Quantity Restrictions (Article 3(5)) and Restriction on Use for a Third Party's Plant (Article 2(1)(12))

From the time when the Commission treated licences as horizontal once licensee and licensor were both manufacturing, it has objected to restrictions on the quantities the licensee might produce.

7.3.12.1 The Black List, Quantity Restrictions (Article 3(5))

Article 2(1)(12) needs to be explained in the context of article 3(5) which blacklists quantity limitations for the reason stated in recital 24:

---

[91] N. 15 above.

(24) Besides the clauses already mentioned, the list of restrictions which render the block exemption inapplicable also includes restrictions regarding the . . . quantities to be manufactured or sold, since they seriously limit the extent to which the licensee can exploit the licensed technology and since quantity restrictions particularly may have the same effect as export bans (article 3(1) and (5))

Article 3(5) prevents the exemption applying where:

(5) the quantity of the licensed products one party may manufacture or sell or the number of operations exploiting the licensed technology he may carry out are subject to limitations, save as provided in Article 1(1)(8) and Article 2(1)(13);

As in recital 23 of the patent regulation, limits of quantity are seen as export bans. The concern about export bans relates to the simplistic attitude the Commission has long taken towards market integration. If the licensee is required to use its best endeavours to supply its own territory, but limited quantitatively in the use of the technology, it may not have much product to export to other Member States. The objection to limitations of quantity applies only to maximum and not to minimum quantities, which are whitelisted by article 2(1)(9).

The express exclusion from the black list of a use licence exempted by article 1(1)(8) is an improvement on the patent regulation, although it may be argued that a licence limited to production for one's own needs is not a quantity restriction, if one's own needs are not limited. The licence may extend only to the amounts needed by the licensee to produce a product or field of use specified.

7.3.12.2 Restriction on Use for a Third Party's Plant (Article 2(1)(12))

The next item of the white list is:

(12) an obligation on the licensee not to use the licensor's technology to construct facilities for third parties; this is without prejudice to the right of the licensee to increase the capacity of his facilities or to set up additional facilities for his own use on normal commercial terms, including the payment of additional royalties;

Recital 24 explains that restrictions on sales prices and on the quantities sold are blacklisted, but that:

(24) . . . This does not apply where a licence is granted for use of the technology in specific production facilities and where both a specific technology is communicated for the setting-up, operation and maintenance of these facilities and the licensee is allowed to increase the capacity of the facilities or to set up further facilities for its own use on normal commercial terms. On the other hand, the licensee may lawfully

be prevented from using the transferred technology to set up facilities for third parties, since the purpose of the agreement is not to permit the licensee to give other producers access to the licensor's technology while it remains secret (article 2(1)(12)).

A licensor is permitted to select his licensees, but is not allowed to limit the quantity they may produce, save under articles 1(1)(8)[92] and 2(1)(13). Even if the licensor builds the necessary equipment for the licensee, the licensee must be permitted to expand its own plant on normal commercial terms save in so far as it is restricted as to field of use under article 2(1)(8). It may be that a different royalty might be charged when the plant is expanded or the same royalty despite the licensee paying for the expansion and the licensor for the original installation, provided that it does not exceed normal commercial terms and that a high royalty is not used to deter expansion.[93] By the time the plant is expanded, the technology may be worth more or less than previously. A market may already have been developed and the risk may be less. The technology may have been improved, but it may be obsolescent. Later licensees may be required to pay more or less, so differences in royalties do not indicate that a particular licensee is being discouraged from producing more. The licensee may be restrained from supplying plant to third parties since recital 24 states that the purpose of the agreement is not to permit the licensee to give other producers access to the licensor's know-how while it remains secret.

Questions may arise about the validity of a site licence. For some products the location of the plant is important; for instance certain beers can be brewed only in places where a particular yeast is in the air. If I license you to use my recipe only at a particular site, does this amount to a restriction on the quantity you can produce? Probably not if there is room for expansion on the site, but the black list may apply if there is not and the licensor refuses to permit the licensee to build for expansion in another suitable site if the licensor asks permission. The licensor should keep records of reasons why permission was not given for the erection of further plant elsewhere and of any alternative suggestions he made.

7.3.13 Second Source (Recital 23 and Article 2(1)(13))

Many manufacturers buy in substantial quantities of components, where there are economies of scale in design that cannot be realized by a single manufacturer. A component manufacturer competing to sell to several

---

[92] Art. 1(1)(8) permits a licence limited to what is needed by the licensee for his own use, but the quantity required may not be limited.

[93] See recital 21 which prevents the reg. applying when rates of royalty are set so as to achieve one of the restrictions in art. 3.

vehicle manufacturers may have a larger turnover over which it can spread
the cost of developing the components. Nevertheless, the history of strikes
against a particular component manufacturer's plants or against transport
facilities may make the vehicle manufacturer's decision to buy only from
the component maker who designed the component risky. It may well
insist that a second source be licensed.

The provision now in article 2(1)(13) was inserted by the know-how
regulation as a specific practice subject to the opposition procedure to
enable the holders of technology in components to assure their customers
of a second source. This gave the Commission a chance to look at some
individual agreements and decide whether they were anti-competitive. No
cases have been published, and it seems that the Commission is now
satisfied that they are not anti-competitive. It may be beginning to return
to the concept of limited licences adopted in its original notice on patent
licences issued in 1962.[94] The insertion of this provision, first in article 4
and now in the white list, may indicate that firms may use this way of
introducing the Commission to provisions considered important in an
industry.

Recital 23, after explaining why in article 3(4) the Commission objects to
customer restrictions, added:

> Article 3 does not apply to cases where the patent or know-how licence is
> granted in order to provide a single customer with a second source of supply.
> In such a case, a prohibition on the second licensee from supplying persons
> other than the customer concerned is an essential condition for the grant of a
> second licence, since the purpose of the transaction is not to create an
> independent supplier in the market. The same applies to limitations on the
> quantities the licensee may supply to the customer concerned.

The arrangement for providing a second source is justified in terms
somewhat similar to the sub-contracting notice.[95] The second source is not
an independent undertaking on the market. Again the Commission looks
ex ante to an ancillary restriction needed to induce dissemination to the
second source.

Article 2(1)(13) whitelists:

> (13) an obligation on the licensee to supply only a limited quantity of the
> licensed product to a particular customer, where the licence was granted
> so that the customer might have a second source of supply inside the
> licensed territory; this provision shall also apply where the customer is

---

[94] See 3.3–3.3.5 above.
[95] [1979] OJ C1/2. This states that where a person's intellectual property rights are used by
his sub-contractor, the sub-contractor should be treated as if it were part of the contractor's
organization, so the agreement does not infringe art. 85(1).

the licensee, and the licence which was granted in order to provide a second source of supply provides that the customer is himself to manufacture the licensed products or to have them manufactured by a subcontractor;

Article 1(1)(13) goes beyond a use licence to a customer to have the products made for its own use in that there may be a maximum quantity restriction. The licence may be granted to the customer itself under Article 1(1)(8) or to a third party to produce for the customer under Article 2(1)(13).

7.3.14 A Licence does not Exhaust the Intellectual Property Right (Recital 11 and Article 2(1)(14))

At 2.1.2 and 2.1.3 above, I explained the doctrine of exhaustion developed by the Community Court in construing articles 30 and 36 of the Treaty. The Court has frequently reiterated paragraph 15 of its judgment in *Centrafarm* v. *Sterling*:[96]

> the exercise, by a patentee, of the right which he enjoys under the legislation of a Member State to prohibit the sale, in that State, of a product protected by the patent which has been marketed in another Member State by the patentee or with his consent is incompatible with the rules of the EEC Treaty concerning the free movements of goods within the Common Market.

As explained at 3.2 to 3.2.5 above, the Commission used to treat a licence as exhausting the intellectual property right. This is the basis on which the patent regulation was drafted. The holder will be able to charge a royalty on granting a licence, but the wording constantly used by the Court in relation to exhaustion refers to marketing in the country of origin by or with the holder's consent and not to direct sales by the licensee for another Member State in the country where the holder tries to enforce its intellectual property rights.

Article 2(1)(14) of the technology transfer regulation implies that the Commission has changed its mind about direct sales by or with the consent of the patentee exhausting the right. It lists:

> (14) a reservation by the licensor of the right to exercise the rights conferred by a patent to oppose the exploitation of the technology by the licensee outside the licensed territory;

This implies that there are some rights left unexhausted in the country of import. Under the national law of some states, in relation to some kinds of

---

[96] *Centrafarm BV* v. *Sterling Drug Inc.* (15/74), [1974] ECR 1147, [1974] 2 CMLR 480, CMR 8246.

intellectual property, there will not be any, but at least it is implied that any national rights there may be are not pre-empted by contrary Community law.

Of course, it for the Court to decide whether to extend the doctrine of exhaustion to indirect sales by a licensee into the territory of another licensee.

This provision is without limit of time. This may surprise those used to a restriction on passive sales outside the territory of the licensee being subject to a limit of five years. Nevertheless, it is not likely to add significantly to the territorial protection that a licensor can grant his licensee, as where the price in the country of sale is higher than that in the licensed territory, a purchaser from the licensee may sell anywhere. Nevertheless, it may be helpful in relation to products that do not pass through dealers.

7.3.15 Rights Where There is a Challenge to the Know-how or Patent, or It Is Alleged that the Patent is not Necessary (Article 2(1)(14) and (15))

These provisions were dealt with in conjunction with no-challenge clauses at 7.3.6.2 above.

7.3.16 Non-competition (Article 3(2))

This provision was considered at 7.3.9 above in connection with some of the whitelisted provisions that qualify it. It differs from the other black- and whitelisted provisions in that it relates not to the licensed technology and products made thereby, but to competing technology and products. Such a clause may, therefore, be anti-competitive if the market be concentrated and the parties actual or potential competitors. It is, therefore somewhat anomalous that it is the only blacklisted provision that can be easily and clearly avoided by careful drafting, substituting an obligation for the licensee to use its best endeavours to exploit the licensed technology and the other provisions analysed at 7.3.9 above.

7.3.17 Price-fixing (Article 3(1))

Article 3(1), like provisions in the former regulations, prevents price-fixing agreements from qualifying for the exemption. It blacklists agreements where:

> (1) one party is restricted in the determination of prices, components of prices or discounts for the licensed products;

Recital 24 states:

> (24) Besides the clauses already mentioned, the list of restrictions which render the block exemption inapplicable also includes restrictions regarding the selling prices of the licensed product or . . .

and then continues to explain the provisions relating to quantity restraints. It does not state why price restrictions should be blacklisted.

Where the parties have market power and are actual or potential competitors, agreement on minimum prices would amount to horizontal cartels, and might have no object other than to restrict production and raise price. Where, however, the relationship is vertical—where the parties are not even potential competitors without the licence—Chicago economists would see nothing wrong with price restrictions: they could be a device for minimizing free-riding. At least a free-rider would be restrained from undercutting the licensee who provides services.

In the United States, however, there is case law confirming that, even in a vertical relationship, maintaining the prices at which the licensee may sell infringes section 1 of the Sherman Act. At the spring meeting of the ABA Antitrust Section in 1996, the heads of both agencies confirmed that they would enforce this law, although Mr Pitofski, Chairman of the Federal Trade Commission, expressed some doubts about forbidding maximum prices above which a licensee should not be allowed to sell. It can be argued that, if the services induced by price protection are wanted only by a few marginal customers, it might be worth while for the licensor to induce them, but would lead to the proliferation of services that most customers do not want.

The US Supreme Court has been loth to find an agreement when a supplier of goods cuts off price-cutting dealers without imposing resale price maintenance.[97] The same ideas might be carried into the licensing context. In Europe, however, a concerted practice is much more easily found.[98]

The question arises whether the blacklisted item should be limited to minimum or fixed prices that raise the level of prices and so restrict demand, or whether it may apply to maximum prices too. Maximum prices may deter exports to countries where prices are higher and are probably blacklisted.

Article 3(1) blacklists price restrictions on either party, so precludes the fixing of prices at which the licensor may sell goods made by the licensed technology to the public.

It may be that since most-favoured licensee terms are whitelisted by

---

[97] *Monsanto Co.* v. *Spray-Rite Service Corp.*, 465 US 752 (1983).

[98] *AEG-Telefunken—Allgemeine Elektricitäts-Gesellschaft AEG Telefunken AG* v. *Commission* (107/82), [1983] ECR 3151, [1984] 3 CMLR 325, CMR 14,018.

article 2(1)(10), the licensor may agree not to sell to the public on terms that a licensee given such an undertaking could not match profitably, but this is far from clear. Sub-paragraph (10) speaks only of 'most favoured licensee', but the list is not exhaustive, so may embrace a promise not to be undersold by the licensor. Setting the prices at which the licensor will sell components or raw materials to the licensee would not be a 'restriction in the determination of prices', but the positive setting of a price between the parties to a sale.

The recommendation of prices is not blacklisted, and probably has no effect on the application of the exemption. In *Pronuptia*,[99] the Court ruled that the recommendation of resale prices that franchisees were free not to charge did not infringe article 85(1), although there is some doubt about extending the precedent beyond franchising, where the franchisor may exercise close control over its franchisees.

7.3.18 Automatic Prolongation (Article 8(3))

Automatic prolongation is no longer blacklisted, as discussed at 5.2.4 above. Article 8(3) includes agreements that are automatically prolonged by the inclusion of improvements within the scope of the regulation, provided that:

> 3. . . . the licensee has the right to refuse such improvements or each party has the right to terminate the agreement at the expiry of the initial term of an agreement and at least every three years thereafter.

7.3.19 Strong Feedback (Article 3(6)), Quantity Limits (Article 3(5)) and Customer Restraints (Article 3(4))

Strong feedback provisions were considered at 7.3.4.4 above and quantity limits and customer restraints at 7.3.8.2. above.

7.3.20 Excessive Territorial Protection (Article 3(3) and (7))

The final items in the black list circumscribe the territorial protection permitted by article 1 and considered at 6.6 above.

7.3.20.1 Impeding Parallel Trade (Recital 17 and Article 3(3))

Recital 17 explains:

> (17) The obligations listed in Article 1 also generally fulfil the other conditions for the application of Article 85(3). Consumers will, as a

---

[99] N. 14 above.

rule, be allowed a fair share of the benefit resulting from the improvement in the supply of goods on the market. To safeguard this effect, however, it is right to exclude from the application of Article 1 cases where the parties agree to refuse to meet demand from users or resellers within their respective territories who would resell for export, or to take other steps to impede parallel imports. The obligations referred to above thus only impose restrictions which are indispensable to the attainment of their objectives.

This is implemented by article 3(3) which blacklists consensual ways of making parallel trade difficult. The exemptions shall not apply where:

(3) one or both of the parties are required without any objectively justified reason:
   (a) to refuse to meet orders from users or resellers in their respective territories who would market products in other territories within the common market;
   (b) to make it difficult for users or resellers to obtain the products from other resellers within the common market, and in particular to exercise intellectual property rights or take measures so as to prevent users or resellers from obtaining outside, or from putting on the market in the licensed territory products which have been lawfully put on the market within the common market by the licensor or with his consent;
   or do so as a result of a concerted practice between them;

This blacklisted item must be without prejudice to the licensor exercising a right reserved in the licence under article 2(1)(14) to sue one licensee for selling outside its territory. Nevertheless, business should beware of article 3(3). Such provisions automatically prevent the application of the exemption even to other provisions, and are likely to result in substantial fines. Even if the original licence comes within the agreement, once one party colludes with the other to impede parallel trade in this way, the exemption will cease to apply. The Court and Commission have easily accepted that unilateral action is concerted in the context of a long-term relationship.

Unfortunately, the provision is worded in the passive voice.[100] Does it apply to a licence granted to L1, when another licensee, L2, requires the licensor to impede parallel trade from L1's customers to L2's territory? Does it apply to L1's licence when L2 requires the licensor to prevent parallel trade from L3's customers? The converse question arises when, at the behest of L2, the licensor discourages a licensee from selling into the territory of another. Is it merely that licence that ceases to be exempt, or do all the licences of the same technology cease to be exempt? To a common lawyer it would seem outrageous if a party to a valid contract might be exposed to penalties and nullity by the action of others. It may

---

[100] This is also true of the French version.

argued that the general principles of legitimate expectations and contractual certainty mitigate against this possibility.

7.3.20.2 Longer Territorial Restraints (Recital 14 and Article 3(7))

Although some territorial protection is permitted by article 1, as considered at 6.6 to 6.6.10 above, recital 14 states that:

> (14) Exemption under Article 85(3) of longer periods of territorial protection for know-how agreements, in particular in order to protect expensive and risky investment or where the parties were not competitors at the date of the grant of the licence, can be granted only by individual decision. On the other hand, parties are free to extend the term of their agreements in order to exploit any subsequent improvement and to provide for the payment of additional royalties. However, in such cases, further periods of territorial protection may be allowed only starting from the date of licensing of the secret improvements in the Community, and by individual decision. Where the research for improvements results in innovations which are distinct from the licensed technology the parties may conclude a new agreement benefitting from an exemption under this Regulation.

The problems of distinguishing continuing improvements and distinct innovations has already been addressed at 6.6.5 to 6.6.7.

The recital is implemented by article 3(7). The regulation does not apply where:

> (7) the licensor is required, albeit in separate agreements or through automatic prolongation of the initial duration of the agreement by the inclusion of any new improvements, for a period exceeding that referred to in Article 1(2) and (3) not to license other undertakings to exploit the licensed technology in the licensed territory, or a party is required for a period exceeding that referred to in Article 1(2) and (3) or Article 1(4) not to exploit the licensed technology in the territory of the other party or of other licensees.

The duration of the territorial protection permitted in article 1(2)–(4) is circumscribed.

8

The Opposition Procedure (Articles 4 and 9)

8.1 INTRODUCTION

For a number of years, the Commission received objections that its group exemptions were too narrow and that, frequently, common kinds of transactions could not be brought within them at all or without commercial loss. It tried to overcome the difficulty by introducing an opposition procedure.[1] Under such a procedure the parties may notify an agreement of the kind covered by the exemption but which contains one or more provisions restricting competition that are not expressly exempted, provided the agreement contains no blacklisted clauses and, in some regulations, that the conditions listed in the regulation are satisfied. The agreement will be exempted under the regulation if the Commission does not oppose the exemption within four months[2] or if it withdraws its opposition at any time. If it maintains its opposition, the agreement falls into the procedure under regulation 17.

One innovation in the know-how regulation was that article 4(2) provided that the procedure applied in one specific situation, now whitelisted by article 2(1)(13).[3] Two provisions are expressly made subject to the opposition procedure in the technology transfer regulation. The provisions in article 4(2)(a)[4] and (b)[5] were described above.

An opposition procedure was first introduced into the regulation for patent licensing: no 2349/84.[6] Recital 25 of that regulation described the opposition procedure as:

---

[1] In relation to this chap. see generally: **C. S. Kerse**, *EC Antitrust Procedure* (3rd edn., London: Sweet and Maxwell, 1994), from para. 2.40 on 85 ff. A supplement to the 3rd edn. is in preparation, but will use the same para. nos.; **James Venit**, 'The Commission's Opposition Procedure—Between the Scylla of *Ultra Vires* and the Charybdis of Perfume: Legal Consequences and Tactical Considerations'. (1985) 22 *CMLRev.* 167; **Eric White**, 'Research and Development Joint Ventures Under EEC Competition Law' (1985) 16 *IIC* 663.

[2] It is 6 months under the earlier regs.   [3] See 7.3.13 above.

[4] Relating to quality specifications and ties, considered at 7.3.5–7.3.5.2 above.

[5] Relating to restrictions on challenging the validity of patents and the secrecy or substantial nature of know-how, considered at 7.3.6.2 above.

[6] 3 regs. of general application other than those for patents and know-how have incorporated an opposition procedure for clauses that are not listed in the reg. but which

a simplified means of benefiting, upon notification, from the legal certainty provided by the block exemption (Article 4). This procedure should at the same time allow the Commission to ensure effective supervision as well as simplifying the administrative control of agreements.

Recital 25 of the technology transfer regulation does not refer so clearly to legal certainty, but does state that qualifying agreements 'may nonetheless be presumed to be eligible for application of article 85(3)'.

The procedure has the advantage of enabling more agreements to qualify for exemption without requiring the Commission to go through the formal process of making individual decisions under article 6 of regulation 17, while at the same time providing a chance for the Commission to check. The procedure also obviates the need for the parties to decide whether ancillary provisions do restrict competition and need exemption. If there is doubt, they can notify.

On the other hand there is an uneasy tension between the automatic exemption granted by a block exemption and the safeguards for parties and complainants provided by regulations 17 and 99/63, which provide for the discretionary grant of individual exemptions. The Commission's power to adopt an opposition procedure is not entirely free from doubt.[7] There is also doubt about the validity of the agreements during the four months after notification or where the Commission has opposed the application of the regulation but might withdraw its opposition at any time. It is not clear what letters written by the Commission under the procedure will amount to 'acts' capable of challenge before the Community Court. In practice it may be difficult to know whether particular clauses are restrictive of competition and may be notified under the opposition procedure and also whether other provisions fall within the black list and prevent the procedure from applying. Few agreements have yet slipped through the procedure.[8]

---

restrict competition: reg. 418/85 relating to collaboration for r & d; reg. 4087/88 which exempts franchising agreements; and reg. 417/85 which incorporates an opposition procedure in relation to the turnover limitation for specialization agreements.

There are also regs. applying in the air and sea transport sectors that incorporate an opposition procedure. In *Ahmed Saeed Flugreisen* v. *Zentrale zur Bekämpfung unlauteren Wettbewerbs* (66/86), [1989] ECR 803, [1990] 4 CMLR 102, [1989] 2 CEC 654, the Court considered the effect of the regs. relating to minimum fares for air transport, but the validity of the opposition procedure was not in issue and was not argued, so at para. 9 the Court left open the question of validity when it proceeded to analyse the effects of the reg.

[7] See 8.5.1 below.

[8] The Commission's annual *Reports on Competition Policy* give the statistics. In most years there are fewer than half a dozen notifications under all the regs. providing an opposition procedure. Many of them are invalid. Sometimes there is a blacklisted provision, sometimes the transaction does not qualify as the kind of agreement to which a group exemption applies. Sometimes there are no provisions that restrict competition.

8.2 NOTIFICATION (ARTICLE 4(1))

8.2.1 Form A/B

Article 4(1) of the technology transfer regulation provides that the exemption applies also to restrictions of competition which are not covered by article 1 or 2, and are not blacklisted, provided that the agreement is notified in accordance with articles 1, 2, and 3 of regulation 3385/94[9] and the Commission does not oppose the exemption within four months.

Regulation 3385/94 is short and provides not only form A/B which lists the information to be provided,[10] but also an explanatory note. Article 4(4)(a) of the know-how regulation used to require express reference to be made to article 4 of the know-how regulation and article 4(4)(b) required that the information be complete and accurate. This has been omitted from the technology transfer regulation. The second requirement is contained in article 3(1) of regulation 3385/94 and heads E and F of the note[11] and clearly persists, but neither the technology transfer regulation nor paragraph 17 of form A/B any longer requires that express reference be made to the opposition procedure and the regulation under which it is being invoked. It is not entirely clear whether this was only for the sake of simplicity or whether the law has been altered. There is, however, little cost in making an express reference to the opposition procedure, so it may be sensible to continue to do so.

Recital 25 of the technology transfer regulation states that:

> (25) Agreements which are not automatically covered by the exemption because they contain provisions that are not expressly exempted by this Regulation and not expressly excluded from exemption, including those listed in Article 4(2), may, in certain circumstances, nonetheless be presumed to be eligible for application of the block exemption. It will be possible for the Commission rapidly to establish whether this is the case on the basis of the information undertakings are obliged to provide under Commission Regulation (EC) No 3385/94. The Commission may

---

[9] [1994] OJ L377/28.

[10] Art. 2(2) of reg. 3385/94 requires 17 copies of the notification and 3 of its annexes to be sent in one of the official languages of the EC. Art. 3(3) is very helpful. It provides that the Commission may waive any items of information where it considers that the information or documents are not necessary.

[11] As **Kerse** suggested, n. 1 above, 2.43 (iii), at 88, this probably does not require the parties to provide more information than is required for a 'normal notification' using the form.

> 'the obligation to supply "complete" information cannot be at large. There must be objective limits, both as regards materiality and what the parties knew or ought reasonably to have known in the circumstances and at the relevant time.'

waive the requirement to supply specific information required in form A/B but which it does not deem necessary. The Commission will generally be content with communication of the text of the agreement and with an estimate, based on directly available data, of the market structure and of the licensee's market share. Such agreements should therefore be deemed to be covered by the exemption provided for in this Regulation where they are notified to the Commission and the Commission does not oppose the application of the exemption within a specified period of time.

This leniency requires a specific waiver by the Commission under article 3(3) of regulation 3385/94. The parties should provide the information indicated in recital 25 of the technology transfer regulation and request a waiver of the rest. The Commission is unlikely to refuse it. It may send letters under regulation 17, article 11, requesting further information, but this will not stop the four months from running.

The shorter notification should make the opposition procedure far more attractive. Finding all the information required to fill in the new version of form A/B is very time-consuming and requires collaboration between people in the firms concerned and expert help from specialized competition lawyers. If information about the market is required only in so far as it is directly available, the task should be considerably simpler and cheaper.

The question arises whether the Commission will be prepared to waive most of the information required also for notification under the opposition procedure of the other group exemptions. There is no need for any legislative action if the Commission wishes to revise its procedure and it is thought that this is likely.

Where the notification is clearly insufficient, the Commission may write back to the parties, but no duty has been expressly imposed on it to do so. If nothing of substance is heard from the Commission, it would be dangerous to assume that the notification is sufficient. In many cases there may be no way for the Commission to know that the notification is incomplete or inaccurate unless it makes a full investigation.

It is important that the notification should be sufficient. If it is, and the Commission does not oppose the exemption within four months,[12] or if it withdraws its opposition at any time—even outside the four-month period[13]—the exemption applies automatically from the date of the

---

[12] Where the information is incomplete in a material respect and the Commission so notifies the parties, the notification dates only from when the supplementary information is provided: art. 4(2) of reg. 3385/94.

Art. 4(5) of reg. 240/96 enables Member States to require the Commission to oppose the regulation on the ground that the application of the reg. would infringe the competition rules of the Treaty.

[13] Art. 4(6) of the technology transfer reg., but see n. 29 below. This applies even when a Member State required the Commission to oppose, provided that the Advisory Committee has been consulted.

notification. If the parties have to amend the agreement, then in the absence of continued opposition by the Commission, the exemption applies from the time the amendments take effect. See 8.3 below. If, however, the notification is incomplete or not in accordance with the facts, the regulation does not apply at all if the ancillary provisions do in fact restrict competition.

If the Commission opposes exemption and does not withdraw its opposition, the effects of the notification are governed by regulation 17.

8.2.2 The Commission's Responses to Notifications

Although the Commission is under no duty to respond to notifications under article 4, it has prepared standard letters for the officials in DG IV responsible for the relevant classes of|products to send in response to notifications under the patent regulation they may also be used|for the technology transfer regulation, but it is too early to say. They provide for senior officials to give their views on the effect of the notification. One standard letter states that the agreement falls within the automatic effect of the group exemption; that the exemption appears from its file to apply from a date specified; and that the Commission's file has been closed. Another standard letter states that specified clauses in the agreement are blacklisted and that the regulation cannot apply, whether or not the opposition procedure has been invoked, that there is no need for the Commission to oppose. It then encourages the parties to delete the blacklisted clauses. The Commission adds that unless it hears within nine weeks that the agreement has been amended it will treat the notification as a request for exemption, although this cannot be envisaged 'in the absence of further and convincing reasons to justify the specified clause under the terms of Article 85(3)'.

Where reference has not been made in the notification to article 4, a third standard letter, less likely to be useful now, is sent stating that, although there are provisions of a kind not listed in article 2 or 3 of the exemption, the Commission is closing its file; that the exemption does not apply; and that it may re-open its file at any time. It does not add that the parties may care to notify under the opposition procedure in order to ensure the validity of their agreement. A fourth standard letter says that such 'grey restrictions'[14] are sufficiently grave not to be acceptable, even had the notification been made under the opposition procedure; that the

[14] A term commonly used to mean provisions which appear in neither the white nor the black list, but which restrict competition.

notification is being treated as having been made under article 85(3), but that an exemption cannot be envisaged in the absence of a convincing justification although, meanwhile, the agreement remains immune from fines.

Where reference has been made to article 4(1), the Commission may state in response to an enquiry that it has not opposed the exemption and that article 4(1) appears to apply, subject to the Commission's power to terminate the exemption under article 7.[15] It may write to the parties that,

---

[15] Since such a letter may well not be very specific and may be approved by only a single Member of the Commission, it would not estop the Commission. See the Community Court's judgment in *Frubo* v. *Commission* (71/74), [1975] ECR 563, [1975] 2 CMLR 123, CMR 8285, paras. 19–20.

> '19. In the letter, the Director-General of Competition, taking note of a specific amendment to the agreement which the applicants were prepared to accept, states that, in his view, the agreement as thus amended, can, notwithstanding the remaining restrictions on completion, qualify for exemption under art. 85(3).
>
> 20. Expressed in these terms, the opinion given could not convey any impression that it committed the Commission; nor, moreover, is the signatory authorised to enter into such a commitment.'

Nevertheless, a letter stating that an agreement does not infringe art. 85(1) because of the undertaking's small share of the market is a matter that may be taken into account by a national court and is useful as long as the market share remains small; see *Guérlain—Procureur de la République* v. *Giry and Guerlain* (253/78, etc.) [1980] ECR 2327, [1981] 2 CMLR 99, CMR 8712:

> '13. Such letters which are based only upon the facts in the Commission's possession, and which reflect the Commission's assessment and bring to an end the procedure of examination by the department of the Commission responsible for this, do not have the effect of preventing national courts, before which the agreements in question are alleged to be incompatible with Article 85, from reaching a different finding as regards the agreements concerned on the basis of the information available to them. Whilst it does not bind the national courts, the opinion transmitted in such letters nevertheless constitutes a factor which the national courts may take into account in examining whether the agreements or conduct in question are in accordance with the provisions or Article 85.'

Such a letter is useful, both before national courts and because the Commission would find it politically difficult to change its mind, unless new relevant facts came to the notice of the Commission before the expiry of the four months. Moreover, just before he ceased to be the member of the Commission in charge of competition, Sir Leon Brittan said in a speech reproduced in **Piet Jan Slot** and **Alison McDonnell** (eds.), *Procedure and Enforcement in EC and US Competition Law: Proceedings of the Leiden Europa Instituut Seminar on User-Friendly Competition Law* (London: Sweet & Maxwell, 1993), 120:

> 'It is true that in theory a comfort letter does not provide complete legal security: the Commission may withdraw it at any time. In fact, however, this is far from true. A formal exemption decision may be revoked if certain conditions specified in Article 8 of Regulation 17, are met. A comfort letter will only be withdrawn in the most extreme cases and only when the conditions of Article 8 are fulfilled.'

Unfortunately one of these conditions is that those circumstances have changed meanwhile, so if the licensed technology is a commercial success and the parties' market share escalates—the very time when they need the protection of an exclusive territory and the associated permitted export bans—the agreement may cease to be enforceable. Moreover, Sir Leon's statement does not bind national courts although they may take it into account. Commercial firms are more likely to be concerned by invalidity than by the possibility of fines when the agreement is vertical and does not involve excessive territorial restraints.

in view of their amendment or further arguments, the Director will not recommend that the Commission oppose. Another standard letter states that the Member of the Commission responsible for competition has taken a decision[16] to oppose the exemption, but that this is without prejudice to the legal consequences of notification, including the immunity from fines provided for by article 15(5)[17] of regulation 17.

Further letters have been prepared under article 11 of regulation 17 asking firms for further information. Where the notification invokes the opposition procedure, the Commission adds that the procedure is activated only if the information given is complete.

8.2.3 Comfort Letters and National Courts

It is thought that a national court, when asked to enforce a contract, should take into account the Commission's views but is not bound to accept them, as the Court ruled in *Guérlain*[18] in relation to a letter stating that, in view of the supplier's small share of the market, the agreement did not infringe article 85(1). The letter informing the parties of the opposition is, of course, in a different category. It is not settled whether a single Member of the Commission, if duly authorized by the Commission as a whole, can do something which would constitute an act binding the Commission; but the validity of that letter is considered at 8.5.4.1 below.

By virtue of article 4(4) of regulation 240/96, where the agreement had already been notified before the regulation came into force on 1 April 1996, the procedure may be initiated by writing to the Commission referring both to article 4 and to the notification. The period of four months starts to run when the Commission receives the notification, but if it is made by registered post (not merely by recorded delivery), it runs from the date shown on the postmark.[19]

8.2.4 Confidentiality (Article 9)

Since it is unlikely that the notification referring to the opposition

---

[16] It is doubted whether a single Member of the Commission has power to take such a decision on his own, but the Commissioners' delegations of power are not public documents, see 8.5.4 n. 53, below.

[17] This presupposes that a notification under art. 4 is also made under reg. 17, which is not entirely clear. See 8.3.1. and 8.4, n. 32 below. Politically speaking, however, it would be impossible for the Commission to impose a fine on the parties after sending such a letter unless the notification was grossly defective. The Court quashed one of the fines in the *Sugar Cases—Re the European Sugar Cartel: Cooperatiëve Vereniging 'Suiker Unie' UA* v. *Commission* (40–48, 50, 54–56, 111, & 113–114/73), [1975] ECR 1663, [1976] 1 CMLR 295, CMR 8334, paras. 555–7, where the Court quashed a fine because the parties might have relied on the notice relating to commercial agents.

[18] N. 15 above; see 8.5.2 n. 46 below, at para. 13.

[19] Art. 4(3) of the technology transfer reg. and art. 4(1) of reg. 3385/94.

procedure is governed by the confidentiality provisions in regulation 17,[20] article 9(1) of the technology transfer regulation provides that information acquired under article 4 shall be used only for the purpose of this regulation. This goes beyond article 20(1) of regulation 17,[21] but citizens are protected from the use of confidential information by Article 214 of the EC Treaty.[21a] On the other hand, regulation 17 provides that the information shall be used only for the purpose of the relevant request or investigation while article 9 permits the use for the purposes of the regulation. If the Commission maintains its opposition, it is likely that regulation 17 would then apply. May information in the notification made under regulation 240/96 be used in deciding whether to forbid the agreement or to grant an individual exemption? It is thought that it should not be, unless the notification was expressly made also under regulation 17, although seventeen copies of a short note incorporating the earlier notification by reference might suffice. See 8.3.1. below.

Article 9(2) of regulation 240/96 provides that:

> The Commission and the authorities of the Member States, their officials and other servants shall not disclose information acquired by them pursuant to this Regulation of the kind covered by the obligation of professional secrecy.

There is no express power in regulation 19/65, under which the regulation exempting technology transfer agreements was made, to provide for confidentiality and there is some doubt whether the Commission has power to extend the obligations of Member States.[22] Article 9 also goes further

---

[20] The relationship between reg. 17 and the group exemptions is unclear; 8.3.1 below. The block exemptions seem to treat notification under the opposition procedure as separate from that under reg. 17, although one can use the same form under both systems.

[21] As stated at 7.3.1 n. 20 above, the guarantee of confidentiality has been confirmed by the Court's judgment in *AKZO Chemie* v. *Commission* (53/85), [1986] ECR 1965, [1987] 1 CMLR 231, CMR 14,318, para. 31. The Court held that a decision by the Commission to give to the complainant information it had obtained at an inspection without giving AKZO an opportunity to protect any business secrets it might contain was illegal, without its being necessary for the Court to ascertain whether the documents did in fact contain any secrets. Consequently, the actual disclosure of the information was also illegal and the Commission was required to do what it could to recover it. The Court thereby prevented the complainant from using in an action in the English courts the information improperly given and discouraged the Commission from violating a general principle, exemplified by art. 214 of the Treaty and art. 20 of reg. 17, which applies throughout the administrative procedure, to take account of undertakings' legitimate interests and, in particular, to maintain their business secrets.

[21a] See *Spanish Banks—Dirección General de Defensa de la Competencia* v. *Asociación Española de Banca Privada* (C–67/91), [1992] ECR I–4785.

[22] The doubt has been reduced since *AKZO Chemie* v. *Commission*, n. 21 above, where the Court based its judgment on a general principle of law protecting the obligation of professional secrecy, merely illustrated by the particular provisions. See *Laos Ludwigshafener Walzmühle Erling KG* v. *Council and Commission* (197–200, 243, 245, & 247/80), [1981] ECR 3211, CMR 8796, grounds 13–16, where the Court ordered that documents improperly obtained should be deleted from the file.

than article 20 of regulation 17 in that, since there is no requirement that a failure to oppose shall be published, there is no need to made an exception for publication in the Official Journal.

It is only information acquired by virtue of the technology transfer regulation and covered by the obligation of professional secrecy that may not be disclosed.[23]

Article 9(3) of regulation 240/96, like article 20(3) of regulation 17, permits the publication of general information and surveys etc.

<h2 style="text-align:center">8.3 OPPOSITION (ARTICLE 4(5))</h2>

Article 4(5) provides that the Commission may oppose the exemption on its own initiative and shall do so at the request of a Member State made within two months of the transmission of the notification to that Member State. It is thought that opposition dates from the time when the parties are notified, and that it should be evidenced in writing. Opposition may affect the validity of a contract, but it will be suggested at 8.5.4.1 below that it may not be an act against which the parties may appeal because the opposition may be withdrawn at any time. Kerse suggests that the Commission can oppose the exemption only on the merits of the agreement for inclusion in the block exemption, although this is expressly provided only in relation to the request of a member state. Once the Commission has opposed the exemption, it is for the parties to justify their agreement, as it is under the procedure of regulation 17.[24]

Paragraph (6) provides that the Commission may withdraw its opposition to the exemption 'at any time'—possibly long after the four months—but if it were raised at the request of a Member State which maintains its request only after consulting the Advisory Committee. Regulation 17, article 10, provides for close and constant liaison with Member States when granting an individual exemption under that regulation. Member States are consulted before automatic group exemptions are adopted, but there is clearly no scope for consultation over a particular agreement when such a regulation applies to it automatically. Under the opposition procedure,

---

[23] See **Kerse**, n. 1 above, 8.19. He explains the difference between this concept and business secrets, as well as obligation under the staff regs. There is no case law from the Community Court on the meaning of the phrase 'obligation of professional secrecy', and Kerse sets out the position in some of the Member States. The obligation in art. 9 may relate to information that comes to an official as such, which the informer has an interest in keeping secret.

[24] See **Kerse**, n. 1 above, at 2.45 and 2.46. He cites *Dutch Book Association: VBVB and VBBB* v. *Commission* (43 & 63/82), [1984] ECR 19, [1985] 1 CMLR 27, CMR 10,730 at para. 52. It is for the parties to suggest amendments that might satisfy the Commission. The Commission may suggest amendments, but it is under no duty to do so.

Member States will still have some influence over the exercise of the Commission's discretion, but less than in the case of an exemption made under regulation 17.[25]

The Commission has no power to make its non-opposition subject to conditions or obligations as it has power to subject the exemption of individual agreements. Kerse suggests that the Commission might, however, suggest amendments to the parties.[26]

Article 4(7) provides that if the opposition is withdrawn because the undertakings have shown that the agreement merits exemption, the exemption dates from the date of the notification. From this it may be inferred that the exemption also dates from notification if no opposition is raised within four months.[27] Problems may well arise about whether the agreement can be enforced against the parties or third parties during the interim period.[28]

Paragraph (8) provides that if the opposition is withdrawn because the parties amended the agreement so that it merits exemption, then the exemption applies from the date on which the amendments take effect.

Unlike exemptions under regulation 17, letting an agreement through under the opposition procedure will be subject to no obligations or conditions. The exemption under the regulation will expire only with the regulation and, on its expiry, one might expect the Commission to renew the block exemption. This may make it slightly more difficult to justify the agreement, although the Commission will still have power to withdraw the exemption by virtue of article 7.

8.3.1 The Relationship with Regulation 17

Article 4(9) provides that:

> If the Commission opposes exemption and the opposition is not withdrawn,

[25] The Commission must take the view that the opposition procedure results in an exemption under the group exemption, and is not a consequence of a decision under reg. 17. Otherwise it would clearly not have power to reduce the role of Member States and the Advisory Committee as provided in reg. 17 of the Council.

[26] The parties might either implement the Commission's suggestions or abrogate those restrictions of competition that the notification under the opposition procedure was made to cure and came within the reg. In the latter event, the Commission would be able to press its suggestions only by withdrawing the group exemption pursuant to art. 7.

[27] A further argument supporting this view may be derived from recital 25 of the reg. exempting patent licences, which refers to the parties 'benefiting, *upon notification*, from the legal certainty provided by the group exemption'. Unfortunately, the French text, *après notification* is less clear and was not included in the *corrigenda* to that reg. published in [1985] JO L280/32.

[28] Actions are unlikely to be brought so soon after the notification, but the validity during the 4 months of provisions that restrict competition may be relevant in later litigation and affect investment decisions meanwhile.

the effects of the notification shall be governed by the provisions of Regulation No 17.

At what point the notification falls into the procedure under regulation 17 is not clear.[29] If the opposition can be withdrawn at any time, the end of the four-month period may not be relevant. The relevant time may not be until the Commission takes an internal decision to initiate proceedings under regulation 17.[29a] National authorities are informed when this happens, as it terminates their power to enforce the EC competition rules under article 9(3) of regulation 17, but the parties are not informed. It is also unclear how far regulation 17 governs the procedure. The Commission may want to keep open the possibility of withdrawing the opposition so that it can finalize an exemption simply without the need for translations and publication in the Official Journal, although this would deprive any party adversely affected by the agreement of its rights under regulation 99/63.

Is notification under the opposition procedure to be treated as notification under regulation 17? Much of the information required to obtain an individual exemption under regulation 17 is to be waived in accordance with recital 25 when the benefit of a block exemption is to be obtained. The same form is required to be used for notifications under both regulation 17 and article 4(1). Does the reference to regulation 17 provide immunity from fines? The Commission's standard letters assume that it does. The question is unlikely to be important, since most people will notify under both regulations. As I suggested at 8.2.4 above, it is not clear that information contained in a notification made only under regulation 240/96 may be used for a decision ordering the parties to desist under article 3 of regulation 17 or only in connection with the group exemption. This may not matter much, as the Commission considers that, even when it cannot use information to establish infringement of the competition rules, it can seek such evidence elsewhere. See 8.4 below.

The notification is, by article 4(1), required to be made in accordance with regulation 3385/94, which was made under regulation 17, but the latter is not expressly mentioned in article 4(1). A senior official, speaking

---

[29] The *vires* for art. 4(9) are not clear. Although the Court considered the effects of an opposition procedure in *Ahmed Saeed Flugreisen* v. *Zentrale zur Bekämpfung unlauteren Wettbewerbs*, n. 6 above, the validity of the procedure was not in issue, and at para. 9 the Court expressly refrained from considering it. At some point, according to art. 4(9), the notification falls into the procedure of reg. 17. Once that point is reached, it must be too late to withdraw the opposition.

[29a] This open-ended situation may be so uncertain and inconvenient that the Commission may decide not to withdraw its opposition outside the 4-month period. If so, it would be desirable to publish its intention.

in a private capacity,[30] stated that notification under the opposition procedure of the know-how regulation counts as having been made under regulation 17. If that is so, it would prevent fines running from the date of notification. The Commission's standard letter for use where it finds a blacklisted clause states that the notification will be treated as an application under article 85(3). Such a letter may well estop the Commission from imposing fines[31] if the information furnished in the notification was sufficient to comply with article 15(5) of regulation 17.

8.4 THE ADVANTAGES AND DISADVANTAGES OF NOTIFYING

The main advantage of notifying under a block exemption is that, in the absence of blacklisted clauses, notification is likely to result in an actual exemption very soon rather than a comfort letter over a year later. The advantage of notifying under regulation 17 is that there clearly is immunity from fines. Information in a notification made under either regulation may not be used for other purposes, as decided in the *Spanish Banks* case.[31a] In practice, information used usually to be furnished under both regulation 17 and the opposition procedure.[32] This is still recommended although it is no longer expressly required.

There is nothing to stop the Commission using its powers to obtain information under articles 11 and 14 of regulation 17 to help it decide whether or not to oppose the exemption. Those provisions refer to carrying out the duties assigned to the Commission under article 89 of the Treaty and the provisions (that is, regulations or directives) adopted under article 87, words general enough to include all the regulations in the competition field. Nevertheless, the Commission has only four months to oppose under the technology transfer regulation and time will continue to run while the undertakings are responding to the letters. So, a request for information, allowing the parties a reasonable time to reply, is hardly practicable, although the Commission may be able to oppose the exemption to allow

---

[30] **Helmut Schröter**, the *Chef de Division* responsible for the drafting of the group exemptions in the Policy Directorate of the Commission's Competition Department, at a conference organized by Longmans in London in May 1985. Other officials have confirmed this, and have stressed the reference to reg. 27 which was aopted under reg. 17. That argument does not seem to me to be sufficient. A reg. may incorporate anything by reference, and one then reads the reg. as if the matter referred to were set out *in extenso* in it. If the officials' views be right, there seems to be no need for art. 4(9) of reg. 240/96.

[31] In *Sugar*, n. 17 above, paras. 555–7, the Court quashed a fine when the parties might have been misled by a careless reading of a Commission notice on commercial agents.

[31a] N. 21a above.

[32] The group exemptions were adopted under Council regs. other than reg. 17. Legally speaking, the systems of exemption are quite separate. To save an administrative burden, however, a single notification requirement is being made under both systems.

for time and then raise its opposition. Firms wanting to pass through the opposition procedure are likely to reply promptly to an informal request for information.

8.5 BENEFITS AND EFFECTS OF THE OPPOSITION PROCEDURE

Exemptions granted pursuant to the opposition procedure will presumably last until the regulation expires at the end of March 2006 and are likely to be prolonged by subsequent legislation, longer than many individual exemptions. The procedure is capable of leading to a considerable number of rapid exemptions and reducing the delays that have occurred under regulation 17. The Commission has found cumbersome its own procedure for granting individual exemptions and although it managed seventeen in 1994, the number was lower in 1995. The opposition procedure may help, at least marginally, to clear the backlog of unclosed files.

It will also save the parties and their advisers from having to guess whether particular ancillary provisions may be found to restrict competition; if in doubt they can use the opposition procedure and ask for waiver for completing form A/B of much of the information most difficult to obtain. Where, however, there is doubt whether the agreement qualifies as a technology transfer agreement, for instance, because the know-how is not very valuable, the opposition procedure does not apply.[33] Provisions that do not significantly restrict competition are clearly necessary to induce pro-competitive investment, or agreements that are not very important may well slip through the opposition procedure and save the parties from the need to distort their agreements so as to bring them within the ambit of the automatic part of the regulation.

[33] In *Boussois/Interpane* [1987] OJ L50/30, [1988] 4 CMLR 124, CMR 10,859, described at 3.1.3 above, the Commission refused to apply the patent reg. to a licence that did not qualify as a patent and know-how licensing agreement. It observed that the know-how dominated the licensed technology; that there were several Member States where there was no patent at all; and that the licensee was not bound to exploit the patents or pay royalties throughout the term of the contract. It continued:

'20. The agreement is therefore not covered by Regulation 2349/84, and until such time as there is a block exemption regulation specifically for pure know-how agreements or mixed agreements in which the know-how component consists of a body of knowledge that is crucial for the exploitation of the licensed technology and not just a factor permitting a better exploitation of the patents, the restrictions of competition involved in the agreement require individual exemption.'

The reasons why the agreement did not come within the concept of a patent licensing agreement are set out in 1.2, text to n. 34 above.

Where it is not clear whether the transaction qualifies under the introductory words of art. 1(1), for instance when the know-how is not very important or is accompanied by a trade mark licence, the parties might consult officials on the basis of a notification under reg. 17, having asked the Commission to waive most of the information about the market.

In the case of more important agreements where restrictions other than those listed in articles 2 and 3 are needed, the Commission may have more difficulty in making up its mind within the requisite four-month period. In that event, however, it can oppose the exemption and can withdraw its opposition at any time, possibly even after the end of the four months.

There are, however, several drawbacks or limitations to the procedure. The *vires* are controversial.[34] It will not always be clear when an agreement is exempt under the opposition procedure, for instance, when it is not clear whether the agreement qualifies as a patent and/or know-how licensing agreement or whether a provision is blacklisted. There are also fewer safeguards for third parties than under regulation 17.

The opposition procedure may, however, be more useful under the technology transfer regulation than under its predecessors. The definition of qualifying licences in the introductory words to article 1(1) is broader and the black list is considerably shorter. Two frequently used provisions have been expressly removed from the black list to article 4(2). Article 5(1)(4) may no longer exclude agreements with important provisions relating to other intellectual property rights such as trade marks, copyright, and software, provided a patent or qualifying know-how is also licensed. Most importantly, the Commission has stated that it will waive much of the information required by regulation 3385/94 when it is not directly available.

8.5.1 *Vires*

The *vires* for the opposition procedure have been questioned.[35] The regulation is made under regulation 19/65, which does not expressly provide for the Commission to have any discretion once the group exemption has been adopted. Article 1(2) of the empowering regulation provides that:

> 2. The regulation shall define the categories of agreements to which it applies and shall specify in particular:
>
> (a) the restrictions or clauses which must not be contained in the agreements;
> (b) the clauses which must be contained in the agreements, or the other conditions which must be satisfied.

---

[34] See 8.5.1 below.

[35] e.g., **Venit**, n. 1 above, 177 ff.; *contra*, **White**, n. 1 above, 697. **Kerse**, n. 1 above, at 2.42 raises doubts, but also gives reasons for thinking that the Court might be sympathetic to the procedure.

As more firms come to rely on agreements where the Commission has not opposed the exemption, an argument based on legitimate expectations becomes stronger and the Court is more likely to confirm the validity of the procedure.

It is not the opposition procedure but articles 1 to 3 that specify the restrictions and clauses that must or may be included and articles 3 and 5 the substantive conditions that must be satisfied for the regulation to apply.

The first issue, then, is whether 'the conditions' referred to in article 1(2)(b) of the empowering regulation are confined to substantive conditions or also include procedural conditions, of the kind provided for by article 4 of the technology transfer regulation, such as notification to the Commission and its not opposing or withdrawing its opposition.

The second objection to the *vires* does not seem to me to be cogent. It is said that the discretion exercised by the Commission is contrary to regulation 17 and reduces the right of the Advisory Committee to be consulted. The Commission's view, however, is that the exemption, if not opposed, is granted by virtue of the technology transfer regulation and not by virtue of regulation 17. The Council has conferred power on the Commission to grant block exemptions and it is not limited by the terms of regulation 17, which is expressed to be only the 'first implementing regulation', implying that there may be others.

The third issue is broader and more important. Regulation 17 envisages that the Commission will exercise considerable discretion under article 85(3). Consequently, safeguards are provided for the rights of the defence and for complainants before an individual exemption can be granted. Article 19(3) of regulation 17 requires that a notice be published in the Official Journal before a decision is adopted so that those adversely affected can protect their position, and regulation 99 gives persons interested and the parties a right to be heard.

The group exemptions granted before 1984 by the Commission apply automatically. Consequently, there is no way that complainants' interests can be protected.[36] The opposition procedure lies uneasily between the concept of the wide discretion and the protection of complainants under regulation 17 and the automatic application of the earlier group exemptions. Although regulation 1017/68 of the Council,[37] which relates to transport by inland waterway, rail, and road, provides in article 12 for something very similar to the opposition procedure, it does not follow that

[36] A complainant has no protection where a group exemption applies automatically, yet the *vires* in reg. 19/65 were upheld by the Court in *Italy* v. *Council and Commission* (32/65), [1966] ECR 389, [1969] CMLR 39, CMR 8048. Does the additional discretion of the Commission under the opposition procedure warrant the protection for complainants of regs. 17 and 99? Since those regs. were adopted, it has become clear that the Commission cannot adopt enough individual decisions and that the resulting uncertainty for business is having serious consequences for the competitiveness of the Community in world markets. It may now be realized that such a high degree of protection for complainants cannot be afforded.

[37] [1968] OJ Spec.Ed. 302. Similarly, some of the later regs. in the transport sector provide for an opposition procedure.

the Commission acting under powers delegated to the Council has as much freedom.

The draft of regulation 17 was considered by the Parliamentary Committee on the Internal Market, and the famous *Deringer Report*[38] questioned the validity of a provision that agreements should become provisionally valid[39] if notified to the Commission in the absence of opposition within four months. Eventually the provision was dropped, probably because of doubts about its propriety.

The third objection to the validity of the opposition procedure is the strongest. Nevertheless, in the past, the Court has taken into account the difficulties the Commission has had in adopting sufficient individual decisions; for instance, in *de Geus* v. *Bosch* it created the concept of provisional validity because of the delays in obtaining decisions from the Commission. Considerable concern has been expressed about the possible invalidity of important provisions in agreements which may deter efficient ways of doing business. It is strongly hoped that the Community Court will eventually uphold the validity of the procedure. In *Italy* v. *Council and Commission*, the Community Court did uphold the validity of regulation 19/65, under which all three technology trnsfer regulations were adopted, partly because:

> 'The need of undertakings to know their legal position with certainty could justify giving priority to the use of this power [to grant group exemptions]. . . (page 404)
> [T]he . . . regulation limits itself to outlining the action which the Commission is to take, while leaving it to the latter to make clear what conditions an agreement must fulfil in order to benefit from an exemption given to a category of agreements (page 406).

It did not limit the conditions to substantive ones, but the possibility of procedural conditions did not arise in that case.

In *Ahmed Saeed Flugreisen* v. *Zentrale zur Bekämpfung unlauteren Wettbewerbs*,[40] the Court considered the effects of the opposition procedure under the airline regulations, but at paragraph 9 it observed that the validity of the regulations had not been questioned in the Court, so it must be taken to have reserved its position on the validity of the opposition procedure.

An alternative strategy would be for the Community Court and

---

[38] Prepared by him as rapporteur for the European Parliament, Doc. 104/1960–1. See **Venit**, n. 1 above, 180.

[39] In 1962, the Community Court accepted the concept of provisional validity for agreements made before reg. 17 came into force in *de Geus* v. *Bosch* (13/61), [1962] ECR 45, [1962] CMLR 1, CMR 8003, but it refused to extend the concept to new agreements, those made after reg. 17 came into force, although the Commission's delays were then well known: *Brasserie de Haecht* v. *Wilkin spouses (No 2)* (48/72), [1973] ECR 77, [1973] CMLR 287, CMR 8170.   [40] N. 6 above.

Commission to hold that fewer agreements infringe article 85(1) of the Treaty.[41]

8.5.2 The Validity of Contracts

The Court refused to extend the doctrine of provisional validity to new agreements in *Brasserie de Haecht II*[42]. It is, therefore, important to know whether an agreement comes within the regulation before commitments are made or resources invested. Even if the *vires* of the opposition procedure are upheld, until the agreement has been notified and the four months have expired, the parties will not know whether the Commission will oppose the exemption. They are unlikely to be enforcing the agreement in a national court at such an early stage, but may have to invest substantial resources to create assets with no other use before knowing whether all or which terms in the agreement will be enforceable. To avoid a possible conflict of decisions between a court and the Commission, a national court should probably adjourn until the four months are up and the Commission should probably not withdraw its opposition after the four-month period.

Nevertheless, if the agreement be unopposed, the actual exemption conferred would be far more satisfactory than the longer periods of uncertainty now common after a notification made under regulation 17 followed by a comfort letter.[43] If the conditions for exemption prevail and the opposition is withdrawn, the exemption will apply from the date of the notification. It is thought that *a fortiori* the exemption will date from notification if the Commission does not oppose.[44] Meanwhile the validity of the agreement seems to be subject to the condition subsequent of the Commission not opposing, or of withdrawing its opposition.

Yet even if the parties hear nothing of substance from the Commission within four months, they may not be sure that the conditions for the

---

[41] As I suggested at 1.1.1, 1.3.1, 1.4–1.4.2.1 above.
[42] N. 39 above.   [43] See 1.1.1 above.
[44] Under English law, a provision that is subject to a condition subsequent operates at once, but is subject to defeasance from the time that the condition is satisfied. The author knows of no general principle of law contrary to this in other Member States. It is derived from ancient Roman law. This theory avoids a legal hiatus between the date of notification and the expiry of the 4 months.
Kerse, n. 1 above, gives this as the more desirable view since it avoids a hiatus, but also suggests a possible alternative to a condition subsequent at 2.47.
> 'Alternatively it is arguable that the exemption attaches at the end of the six months, and, subject to the condition precedent that no opposition has been raised, relates back to the date of notification. The position, although not unreasonable in principle, demands a lot from the silence of the regulations, and does not avoid a legal hiatus, at least temporarily. It will be appreciated that the question of the legal effect of notification may be significant for the status of the acts of raising, maintaining and withdrawing opposition described below.'

application of the procedure have been or will be met. This issue will be for national courts asked to enforce the agreement to decide.[45] The precise ambit of some of the provisions in article 3 is unclear although the uncertainties under the earlier regulations have been reduced. ıIt is now clear that some of the provisos to blacklisted provisions in the know-how regulation are whitelisted. Article 3(4) blacklists customer restrictions only within the same technical field of use or within the same product market, and it is not clear what the first expression means, nor is it easy to be sure that one has applied the second criterion correctly. This provision now applies, however, only if the parties were competing manufacturers at the date of the licence, so will create difficulty less often.

Since the ambit of article 3 is sometimes unclear there would be some risk in relying on the absence of a substantive response from the Commission or on a letter from a Director in DG IV stating that article 4(1) appears to apply to indicate that the agreement is valid.[46] This uncertainty may continue throughout the duration of the agreement.

If the opposition procedure is held to be *ultra vires*, or if a national court holds that the conditions for its application do not prevail, it is hoped that the validity of some agreements may be saved by holding that ancillary restrictions on conduct do not restrict competition and that the exemption of article 1 applies automatically.[47] Alternatively, it may be held that the whole agreement does not restrict competition.[48]

It is not clear whether provisions in the agreement that restrict competition should be enforced once the Commission opposes the exemption. It may later be persuaded to withdraw its opposition without any amendment to the agreement being necessary. To avoid the risk of conflicting decisions, a national court should probably adjourn proceedings

---

[45] In relation to the application of the group exemption granted by reg. 67/67, the Community Court has frequently replied to questions from national courts, thereby implying that they are required to decide whether a group exemption applies, e.g. *Fonderies Roubaix-Wattrelos* v. *Fonderies Roux* (63/75), [1976] ECR 111, [1976] 1 CMLR 538, CMR 8341; *De Bloos* v. *Bouyer* (59/77), [1977] ECR 2359, [1978] 1 CMLR 511, CMR 8444; *De Norre* v. *NV 'Brouwerij Concordia'* (47/76), [1977] ECR 65, [1977] 1 CMLR 378, CMR 8386; *Hydrotherm* v. *Andreoli* (170/83), [1984] ECR 2999, [1985] 3 CMLR 224, CMR 14,112; and *Delimitis (Stergios)* v. *Henniger Bräu AG* (C–234/89), [1991] ECR I–935, [1992] 5 CMLR 210, [1992] 2 CEC 530.

[46] In the *Perfume* cases, e.g., *Guérlain*, n. 15 above, the Community Court ruled that a national court was not bound by a comfort letter, but might take it into account when deciding whether art. 85(1) was infringed. In *De Bloos* v. *Bouyer*, n. 45 above, it ruled that in the light of additional facts the national court should ignore a comfort letter.

[47] The opposition procedure is useful only if in an agreement that would otherwise fall within the reg. there is an unlisted restriction of competition. Where a restriction of conduct does not restrict competition there is no need to use the opposition procedure. It is convenient to notify under the opposition procedure when one is unsure whether a restriction on conduct would be considered by a national court or the Commission necessary to make viable some pro-competitive transaction.

[48] See 1.4–1.4.2.1, 2.2–2.2.5, and 3.1.8 above.

until the Commission decides one way or another. If the Commission eventually withdraws its opposition, the national court would, presumably, still have to be satisfied that the conditions for the opposition procedure are satisfied. Perhaps it might go as far as to make that assessment before adjourning to enable the Commission to decide whether to withdraw its opposition.

8.5.3 The Competence of National Authorities under Regulation 17

The Commission has been considering decentralization of the enforcement of the competition rules and has tried to persuade competent authorities to take a more active role.[49] Nevertheless, national authorities have in practice not been enforcing them. There are many disincentives to their doing so: they have no competence to grant exemptions under article 85(3); many agreements extend beyond a single Member State; and the initiation of procedure by the Commission would terminate their competence to apply articles 85(1) and 86.[50] The opposition procedure was clearly adopted under the group exemption, so does not terminate the authority of national authorities until a procedure is initiated under regulation 17, whenever that may be.[51]

8.5.4 What Letters of the Commission are Capable of Challenge before the Community Court

If the opposition procedure is to operate simply and expeditiously, it is important that few appeals, if any, be taken to the Community Court. Such letters of the Commission as amount to acts within the meaning of article 173[52] must be issued by the Commission as a whole acting collegiately,[53] or

---

[49] Most Member States have adopted competition rules based on arts. 85 and 86, and some national authorities, especially the Italian, are enforcing these.

The national authorities' conference in 1994 proposed, however, that national authorities should enforce arts. 85 and 86 where:
  (1) the case essentially involves a single Member State;
  (2) there is a clear infringement of Community rules, and an agreement has no chance of being exempted by the Commission by virtue of art. 85(3); and
  (3) the national authority can impose effective remedies.
See also, the short *Report on Competition Policy for 1994* (Brussels: EC Commission, 1995), 23, which sets out the first two points.

[50] The Italian competition authority, the Autorità Garante della Concorrenza e del Mercato, will publish the papers on this topic given at a splendid conference it organized in Nov. 1995. [51] 8.3.1 above.

[52] Before judgment was given in *IBM* v. *Commission* (60/81), [1981] ECR 2639, [1981] 3 CMLR 635, CMR 8708, I discussed the issues in **Valentine Korah**, 'Comfort Letters—Reflections on the Perfume Cases' (1981) 6 *ELRev.* 14, 30–4.

[53] Nevertheless in 3 cases, *Demo-Studio Schmidt* v. *Commission* (210/81), [1983] ECR 3045, [1984] 1 CMLR 63, CMR 14,009; *Fediol* v. *Commission* (191/82), [1983] ECR 2913,

by a single Commissioner (or official of the Commission) acting under a delegation of powers to him made by the Commission acting collegiately,[54] and state the reasons on which they are based with sufficient clarity to enable the Community Court to decide whether they are valid.[55] Within two months of being notified of an act, Member States and the person to whom it is addressed or to whom it is of direct and individual concern may petition the Court under article 173.

In *IBM* v. *Commission*,[56] IBM claimed that the statement of objections sent to it was an act against which an appeal lay, but the Court rejected the petition as inadmissible.

> 9. In order to ascertain whether the measures in question are acts within the meaning of Article 173 it is necessary, therefore, to look to their substance. According to the consistent case-law of the Court any measure the legal effects of which are binding on, and capable of affecting the interests of, the applicant by bringing about a distinct change in his legal position is an act or decision which may be the subject of an action under Article 173 for a declaration that it is void. However, the form in which such acts or decisions are cast is, in principle, immaterial as regards the question whether they are open to challenge under that Article.
>
> 10. In the case of acts or decisions adopted by a procedure involving several stages, in particular where they are the culmination of an internal procedure, it is clear from the case-law that in principle an act is open to review only if it is a measure *definitively laying down the position of the Commission or the Council* on the conclusion of that procedure, and not a provisional measure intended to pave the way for the final decision [my emphasis].

The Court protected the Commission from judicial review at an early stage of its proceedings, although, at ground 23, it left open the possibility of early review 'in exceptional circumstances, where the measures

---

[1984] 3 CMLR 244, CMR 14,013 (a case on anti-dumping duty); and *Comité des Industries Cinématographiques des Communautés Européennes* v. *Commission* (298/83), [1985] ECR 1105, [1986] 1 CMLR 486, CMR 14,157, where a complainant alleged that the Commission had failed to pursue a complaint contrary to art. 175, the Court held that the appeal was admissible although the letters stating the Commission's final decision not to act appear to have been written by senior officials. It is not stated that they were authorized by the Members of the Commission and the Commission did not challenge that aspect of admissibility. Where the Commission does not contest the admissibility of the petition, the Court will not know whether letters written by senior officials have been authorized by the Members of the Commission acting collegiately. If they have not been, the Commission has a *de facto* power to decide at the time of the litigation whether to adopt such letters—an anomalous situation.

[54] Cf. *AKZO* v. *Commission*, n. 21 above, paras. 30–37.
[55] As required by art. 190. This duty is not very extensive.
[56] N. 52 above.

concerned lack even the appearance of legality'.[57] Similar reasoning may lead the Court to refuse jurisdiction over many of the letters written in pursuance of the opposition procedure.

8.5.4.1 Opposing the Exemption

Kerse argues at 2.45 that when the Commission opposes the application of the exemption, it should provide evidence of this in writing. Although a letter from the Commission is clearly capable of being an act,[58] it is thought that one opposing the exemption may not be a final 'measure definitively laying down the position of the Commission'.[58a] The Commission may still decide to withdraw the opposition or it may grant an individual exemption under regulation 17.[59] Whether the Member of the Commission responsible for competition policy can oppose the exemption on his own[60] may not come before the Community Court directly under article 173, but could be decided under Article 177 if one of the parties was enforcing the agreement in a national court.

In my view the parties probably cannot appeal from a letter opposing the exemption, as this does not exclude the possibility of exemption. The opposition may later be withdrawn, or an exemption given under regulation 17, although the opposition may delay an exemption indefinitely.

---

[57] The allegations made against the points of objection by IBM were very serious.

[58] *Cimenteries* v. *Commission* (8–11/66), [1967] ECR 75, [1967] CMLR 77, CMR 8052, a letter under reg. 17, art. 15(6), exposing the recipient to liability to fines.

[58a] Opposing the exemption may, however, affect the validity of the contract. A formal decision under reg. 17 is far less easy to obtain than a decision not to oppose and its preparation may take years. The position of the parties is changed.

[59] Contrast analysis of withdrawing the opposition at 8.5.4.2, below.

My former colleague, **John Usher**, has suggested orally that *Fediol*, n. 53 above, shows that a letter concluding one stage of an anti-dumping procedure may be challenged by a complainant where, otherwise, he would have no way of bringing before the Community Court issues on the procedural guarantees of his legitimate interest in having anti-dumping procedures initiated. In that case, however, the letter indicated that the Commission would not initiate such proceedings. This precludes any subsequent proceedings, and so would end the matter unless a challenge were admissible. *Fediol* is welcome in that it protects the rights of complainants, but is difficult to reconcile with cases under the competition rules such as *Lord Bethell* v. *Commission* (246/81), [1982] ECR 2277, [1982] 3 CMLR 300, CMR 8858, or *GEMA* v. *Commission* (125/78), [1979] ECR 3173, [1980] 2 CMLR 177, CMR 8568. In *BAT and Reynolds* v. *Commission*, the *Philip Morris* case (142 & 156/85), [1987] ECR 4487, [1988] 4 CMLR 24 at 32–3; comment by **Paul Lasok** and **Valentine Korah**, (1988) 25 *CMLRev*. 333, at 339. Mancini AG, however, advised the Court to concentrate on general principles and the wording of the legislation, rather than on its earlier case law. See also *AKZO* v. *Commission*, n. 21 above.

[60] See 8.5.4, text to nn. 53 and 54 above. The question turns on whether or not authorization has properly been given to him by the Commission acting as a collegiate body. It is believed it has, but those delegations are not published.

8.5.4.2 *Withdrawing its Opposition*

Unlike a letter opposing the exemption, one withdrawing the opposition definitively defines the Commission's position and may well amount to an act. When the letter is written after the parties have amended the agreement to meet objections from the Commission, it renders valid any anti-competitive provisions not listed in articles 1 to 3 of the regulation even where those provisions were previously unenforceable. Consequently, it clearly affects the legal position of the parties.

When the opposition is withdrawn because the parties have convinced the Commission that the agreement originally notified merits exemption, without amendment, the agreement will be valid provided it complies with the regulation. Has the letter, however, 'affected the interests of the [parties] by bringing about a distinct change in [their] legal position'? If, as seems likely but is not certain,[61] the agreement always was valid subject to a condition subsequent of the exemption not being opposed or the opposition withdrawn, does the failure of the condition subsequent change the parties' legal position? A letter rendering the parties liable to be fined under article 15(6) of regulation 17 has been held to be an act[62] adversely[63] affecting the parties' interests, although the fines would be imposed only if the Commission exercised its discretion against them. In my view the operation of a condition subsequent may also affect the parties' interests. If one of the parties wishes to renegotiate the agreement, he can probably appeal to the Community Court. A complainant is more likely to be adversely affected by the contract becoming valid. He can appeal against a decision addressed to a third party only if it is of direct and individual concern to him.[64] Once he hears of the act, he has two months in which to appeal.

8.5.4.3 *Maintaining its Opposition to the Exemption*

A letter stating that the Commission intends to maintain its opposition and deal with the agreement under regulation 17 may not always be sent to the parties. There is no express requirement that the Commission should inform the parties. If a letter be sent, it may be argued that it is not an act at all: that it has no legal effects on the validity of the agreement or otherwise because it simply repeats the earlier letter setting out the Commission's opposition to exemption. It may indicate that a final

---

[61] See 8.5.2, text after n. 43 above.   [62] See 8.5.4.1, n. 58 above.

[63] In *IBM*, n. 52 above, the Court dropped the qualification 'adversely', which it had implied in *Cimenteries*, n. 58 above, at 91, where it admitted an appeal from a letter depriving the parties of the immunity from fines conferred by art. 15(5) of reg. 17.

[64] See 8.5.4.4 below on the rights of a complainant to appeal.

decision will be delayed, but it is thought that it is not a final act. The Commission may eventually withdraw its opposition, with retroactive effect, or it may grant an individual exemption. Such a measure does not seem to me to be final and definitive. It is almost certainly not subject to appeal.

8.5.4.4 Not Opposing the Exemption

When an exemption is given individually under regulation 17 third parties' rights are protected: the Community Court has held in *Metro* v. *Commission*[65] that the complainant before the Commission is entitled to appeal to it against a decision addressed to its supplier which is of immediate and individual concern to it. Such a complainant is entitled under regulation 99, article 6, to be informed of the reasons why the Commission has rejected the complaint. Where the group exemption is not opposed, however, no decision will be addressed to the parties. It is the regulation that exempts the agreement rather than the decision not to oppose. The parties to the agreement are unlikely to want to renegotiate and appeal within the two-month period allowed by article 173 of the Treaty, even if failure to oppose does amount to an act.

A third party adversely affected by the agreement is more likely to want to appeal. There is no provision for publicity enabling it to object. Even if a third party hears of the Commission's proceedings, a letter stating that the Commission does not oppose the exemption can hardly amount to an act against which it could complain. The satisfaction of the condition is unlikely to be so treated.

It was suggested[66] some time ago that an action for failure to act might be brought under article 175. One difficulty encountered by this view is that article 175 permits an appeal only by the person who claims that the act should have been addressed to him. Unlike article 173, article 175 does not protect those to whom the act is of direct and individual concern. The Court has more recently been interpreting article 175 more widely to admit an action when the act not adopted would have been of direct and individual concern to the applicant.[66a]

---

[65] (26/76), [1977] ECR 1875, [1978] 2 CMLR 1, CMR 8435.

[66] **Dieter Hoffmann**, when still working in the intellectual property division of Directorate D of the Competition Department, stated this as his personal view, but this point was not mentioned in the summary published in his article in (1985) 82 *Law Soc. Gaz.* 23 Jan. at 206.

[66a] See *Empressa Nacional de Uranio SA* v. *Commission* (C–107/91), [1993] ECR I–559. The judgment was given under art. 148 of the Euratom Treaty, but at para. 10 the Court observed that the language was identical with that of art. 175 of the EC Treaty and it relied on case law developed under art. 175.

A second difficulty is that the Commission has obviously been given a discretion, so is under no clear duty to oppose the exemption. The judgment in *Lord Bethell* v. *Commission*[67] denying jurisdiction under article 175 has probably been overruled.

8.5.4.5 *Conclusions on the Possibilities of Appeal*

There is virtually no protection for complainants from the opposition procedure,[68] although a third party may persuade the Commission to withdraw the exemption under article 7. A potential complainant may not know of the notification of agreements under article 4 and, even if it finds out, it almost certainly has no *locus standi* before the Community Court under article 173 or 175 of the EEC Treaty. This might be a reason for the Court to hold that the whole opposition procedure is invalid, which I would regret. As time passes and more undertakings rely on the opposition procedure, however, the Community Court is likely to be influenced by their legitimate expectations and less likely to hold that a useful procedure is invalid. It might do this on a reference from a national court under article 177 when the national court asks questions about the validity and construction of the regulation in order to decide whether it may enforce a contractual provision.

The parties to the contract can almost certainly not appeal against the opposition itself, and four months after notifying it is unlikely either would want to renegotiate and appeal against a failure to oppose.

8.6 CONCLUSION ON THE OPPOSITION PROCEDURE GENERALLY

Personally, I welcome the Commission's attempt to validate agreements that contain terms that do not come within articles 1 to 3. It gives businessmen greater flexibility in negotiating their agreements and saves the parties from having to take a view on whether ancillary clauses infringe article 85(1), when the Court and Commission are applying contrasting standards. It is more likely to be useful than under the earlier regulations. The black list has been reduced. The *vires* are becoming less controversial. Nevertheless, it is no panacea. If the procedure is valid, national courts will have to decide whether there are any clauses that are not listed, and whether there are any blacklisted clauses that prevent the procedure from applying. Legal certainty will be limited.

Some businessmen or their advisers may prefer to rely on the developing

---

[67] N. 59 above.   [68] Or, indeed, from the automatic operation of the reg.

case law of the Court and argue in a national court that the agreement does not restrict competition. They may not notify. Nevertheless, many agreements between firms with small market shares which contain minor restrictions of competition may slip through the procedure and it enables the Commission also to let more important agreements qualify under the regulation.

9

Miscellaneous Provisions

As is required by article 7 of regulation 19/65, the Commission has taken power to withdraw the benefit of the group exemption by individual decision when it considers that a technology transfer agreement has effects that are incompatible with article 85(3) of the Treaty and, in particular, where one or more of four circumstances prevails. The power to withdraw is general and not limited to the listed circumstances.

The introductory words to article 7 provide that:

> The Commission may withdraw the benefit of this Regulation, pursuant to Article 7 of Regulation No 19/65/EEC, where it finds in a particular case that an agreement exempted by this Regulation nevertheless has certain effects which are incompatible with the conditions laid down in Article 85(3) of the Treaty, and in particular where:

There follows a list of four circumstances, which are considered from 9.1.3 to 9.1.3.4 below.

9.1.1 The Effect of Withdrawing the Group Exemption on Article 85

It is almost certain that the exemption from the prohibition of article 85(1) continues to be valid unless and until the Commission takes a formal decision under article 7. In *Tetra Pak I*,[1] the Commission persuaded Tetra Pak to abrogate the exclusivity in a licence it had acquired by merger. Otherwise it would have withdrawn the group exemption for patent licences. It implied that the exemption was valid meanwhile. This is consistent with the words that 'the Commission may withdraw the benefit of the exemption', which imply that the benefit existed to be withdrawn. In *Delimitis*,[2] the Court referred to the need for a national court to avoid decisions conflicting with those of the Commission. Until the exemption is

---

[1] *Tetra Pak I (BTG Licence)* [1988] OJ L272/27, [1990] 4 CMLR 47, CMR 11,015, paras. 28 and 53. Confirmed on appeal on another ground, *Tetra Pak Rausing SA* v. *Commission* (T–51/89), [1990] ECR II–309, [1991] 4 CMLR 334, [1990] 2 CEC 409, paras 21–4.

[2] *Delimitis (Stergios)* v. *Henniger Bräu AG* (C–234/89), [1991] ECR I–935, [1992] 5 CMLR 210, [1992] 2 CEC 530, para. 49.

withdrawn, or the parties have agreed to abrogate the offending parts of an agreement, there is no danger of a conflicting decision.

In *Decca Navigator System*,[3] however, the Commission alleged that a distribution agreement was automatically excluded from the group exemption granted by regulation 67/67 because the agreement had effects incompatible with article 85(3). This must be wrong, and it is hoped that the precedent will not be followed. There would be no point in providing for withdrawal of the exemption when the criteria of article 85(3) are not satisfied if the exemption automatically does not apply in those circumstances. More important, automatic non-application would take away the legal certainty, which it is the function of group exemptions to confer.

The provisions for withdrawal in the various block exemptions are less worrying for business than the provisions for automatic termination or exclusion. They have been exercised by formal decision only once, in *Langnese*,[4] although it is said that on occasion parties have altered their conduct or agreements as did Tetra Pak, to avoid the adoption of such a decision under various group exemptions. Nevertheless, the power is important as it enables the Commission to bring pressure on the parties to amend their agreements or otherwise reduce the anti-competitive effects of their agreements, as it did in *Tetra Pak I*.

The Commission can exercise its power to withdraw the group exemption only by decision, since withdrawal would affect the rights of the parties.[5] It would invalidate the restrictive provisions in the agreement and might render the parties liable to fines for the future.[6] Consequently, the

---

[3] *Decca Navigator System* [1989] OJ L43/27, [1990] 4 CMLR 627, [1989] 1 CEC 137, para. 127.

[4] *Langnese-Iglo GmbH & Co. KG* (93/406/EC), [1993] OJ L183/19, [1994] 4 CMLR 51, [1993] 2 CEC 2123, paras. 115–48, confirmed on this point by the Court of First Instance, *Langnese-Iglo GmbH & Co. KG* v. *Commission* (T–7/93), [1995] 2 CEC 217, [1995] 5 CMLR 602, paras. 173–87.

[5] *Cimenteries—Re Noordwijks Cement Accoord: Cimenteries CBR Cementbedrijven NV* v. *Commission* (8–11/66), [1967] ECR 75, [1967] CMLR 77, CMR 8052, at 91 (ECR).

[6] Where the agreement has been notified under reg. 3385/94 [1994] OJ L377/28, the immunity from fines conferred by reg. 17, art. 15(5), would continue unless the Commission were to issue a decision under art. 15(6) withdrawing immunity from fines at the same time. It may be expected to do so. Fines would probably be payable only for conduct after receipt of the notice under art. 15(6). A literal construction of art. 15(6) would enable the Commission to impose fines for the earlier period, but this has never been established. Not only would it attract adverse political reactions, it would probably defeat legitimate expectations and be disproportionate, contrary to two general principles of Community law.

In *Langnese*, n. 4 above, the Commission formally withdrew the group exemption, and ordered Langnese to desist from tying retailers by exclusive purchasing obligations. So it did not need to use its powers to withdraw the immunity from fines. The Court of First Instance at paras. 205–11, however, held that a block exemption could be withdrawn only from existing and not future agreements with similar provisions.

'208. Regulation 1984/83, being a measure of general application makes available to undertakings a bloc exemption for certain exclusive purchasing agreements which

decision would have to state the grounds on which it is based in accordance with article 190 of the Treaty, would probably have to be made by the Members of the Commission acting collegiately, and would be subject to appeal to the Community Court.[7]

If the Commission considers exercising the power of withdrawal in a particular case, it has to give each party an opportunity to make its views known—that requirement must apply to each of the licensees individually. There would be a chance to argue with the Commission that its tentative objections were wrong, or to change the agreement so as to remove the cause for the Commission's concern. The first sentence of recital 26 of the technology regulation is purely formal and gives no guidance as to how the Commission will act.

9.1.2 The Effect of the Group Exemption on Article 86

Acquiring a firm that enjoys the benefit of an exclusive agreement that comes within a group exemption can amount to the abusive exploitation of a dominant position contrary to article 86. Entering into such an agreement may also infringe article 86. In *Tetra Pak I,*[8] the Commission considered that Tetra Pak had strengthened its position in the market through acquiring the exclusive licensee of the only technology that competed with its own. The licence fell within the terms of the group exemption for patent licensing and the Commission would have withdrawn the exemption had Tetra Pak not renounced the exclusive element. The Commission stated that until Tetra Pak did so, the licence infringed article 86, because the agreement had effects that did not merit exemption. The Commission took the view that the exclusive licence obtained through an acquisition infringed article 86, but not article 85, until the exemption was withdrawn. That view was upheld by the Court of First Instance.[9]

> satisfy in principle the conditions laid down by Article 85(3). According to the hierarchy of legal rules, the Commission is not empowered, by means of an individual decision, to restrict or limit the legal effects of a legislative measure, unless the latter expressly provides a legal basis for that purpose. Although Article 14 of Regulation 1984/83 confers on the Commission power to withdraw the benefit of the regulation if it finds that, in a particular case, an exempted agreement nevertheless has certain effects which are incompatible with the conditions set out in Article 85(3) of the Treaty, Article 14 does not provide any legal basis for the benefit of the block exemption to be withheld from future agreements.'

[7] See **Cimenteries**, n. 5 and 8.5.4 ff. above. The same criteria as to what is an act discussed there are relevant here.

[8] N. 1 above, at paras. 27 and 45.

[9] N. 1 above. The decision may be correct on the facts, since it could be said that the abuse consisted in acquiring a firm with the exclusive right to the only technology capable of competing with its own. The exclusive licence granted to the target firm was not anti-competitive: the acquisition of the target by a competitor was. It might have been more

In *Decca Navigator Systems*,[10] too, the Commission said:

> 122. As all the agreements in question arise from an abuse of a dominant
> position in violation of Article 86, they cannot benefit from the application of
> Article 85(3), either by individual exemption or by general exemption
> provided by regulation.

The view that article 86 may be infringed by an agreement to which a
group exemption applies gives rise to concern as it reduces the legal
certainty attributable to a group exemption. Now that the threshold for
dominance is low and firms with no power over price have been found to
be dominant, the uncertainty may affect many firms. Moreover, the
Commission, with some support from the Court of First Instance, has been
developing a concept of joint dominance to catch abuses where several
firms adopt, or may be expected to adopt, similar conduct in concentrated
markets.[11]

---

difficult for the Commission to find an infringement of art. 86 had the original exclusive
licence that qualified under the patent reg. been granted directly to Tetra Pak, which was
exploiting the only other technology for equipment for making sterilized cartons for long-life
milk, which must last for several months. Only if it were already dominant would art. 86 be
infringed.

[10] N. 3 above.

[11] In *Italian Flat Glass* [1989] OJ L33/44, [1989] 1 CEC 2077, the Commission found the
parties guilty of collusion contrary to art. 85, but added that the firms had also abused a
jointly held dominant position because they presented a common face to the outside world.
The Court of First Instance did not have to address the question, since it had found that the
Commission's market analysis was defective and that joint dominance over the Italian market
had not been established. The Court of First Instance stated on appeal in *Società Italiana
Vetro SpA* v. *Commission* (T–68, 77, & 78/89), [1992] ECR II–1403, [1992] 5 CMLR 302,
[1992] 2 CEC 33, however, that:

> '358. . . . There is nothing, in principle, to prevent two or more independent
> economic entities from being, on a specific market, united by such economic links that,
> by virtue of that fact, together they hold a dominant position *vis-à-vis* the other
> operators on the same market. This could be the case, for example, where two or more
> independent undertakings jointly have, through agreements or licences, a technological
> lead affording them the power to behave to an appreciable extent independently of
> their competitors, their customers and ultimately of their consumers'.

The Commission has on several occasions considered whether a merger would lead to joint
dominance, for instance in *Nestlé/Perrier*, M190, [1992] OJ L356/1, [1993] 4 CMLR M17,
[1993] 1 CEC 2018, and *Mannesmann/Vallourec/Ilva* (94/208/EC), [1994] OJ L102/15, [1994] 1
CEC 2136.

Earlier in *Ahmed Saeed Flugreisen* v. *Zentrale zur Bekämpfung unlauteren Wettbewerbs*
(66/86), [1989] ECR 803, [1990] 4 CMLR 102, [1989] 2 CEC 654, however, the Court did not
accept the concept of joint dominance although, by virtue of reg. 41 there was still no
implementing reg. for air transport where the destination was outside the common market. In
that situation, agreements enjoy provisional validity, so it must have been tempting to apply
art. 86. Nevertheless, the Court said:

> '34. The second problem raised by the second preliminary question is whether the
> application of a tariff [agreed by the carriers and approved by the national authorities]
> may in principle constitute an abuse of a dominant position where it is the result of an

9.1.3 Examples

The introductory words of article 7 are followed by four examples. There were eight examples in the know-how regulation.[12] The reduction in number may not be important since it is the general words that matter, but it may indicate that the Commission is likely to intervene in more limited circumstances.

9.1.3.1 The Lack of Effective Competition on the Demand Side (Article 7(1))

Under the drafts of the technology transfer regulation, it was not to apply where the licensee or both parties had market power. This was defined in various ways in different drafts, mostly unpublished, and aroused a furore. Since the finder of fact, whether the Commission or a national court asked to enforce a contract, has considerable discretion in selecting the relevant market, it is impossible to obtain reliable advice on a firm's market share. So, if enacted, the provision would have led to great uncertainty.

Eventually, the Commission decided to drop the limitations of market share to the application of the regulation in article 1(5) and (6), but it has added twelve words to article 7(2) of the know-how regulation to specify the licensee's 40 per cent share of the market—the test given in article 1(5) of the latest unpublished draft of the regulation that circulated widely.

The Commission may withdraw the exemption where:

> (1) the effect of the agreement is to prevent the licensed products from being exposed to effective competition in the licensed territory from identical

agreement between two undertakings which, itself, is capable of falling within the prohibition set out in Article 85(1).'
After stating that art. 85 applies only to an agreement between independent undertakings, the Court went on to say:
> '37. Those considerations do not exclude the case where an agreement between two or more undertakings which simply constitutes the formal measure setting the seal on an economic reality characterized by the fact that an undertaking in a dominant position has succeeded in having the tariffs in question applied by other undertakings. In such a case, the possibility that Articles 85 and 86 may both be applicable cannot be ruled out. Moreover, the new Council regulations are based on the same interpretation of arts. 85 and 86 in so far as they provide that Article 86 may be applicable to a concerted practice which was initially granted a block exemption . . . or an individual exemption under the objections procedure.'

This idea of the Court seems to be that a dominant firm infringes art. 86 if it exploits its dominance by bullying a firm with less market power to increase the prices it charges, not that joint action by oligopolists infringes art. 86.

[12] Reg. 240/96 does not mention the effects of an arbitration award, the right of the licensor to terminate the licence if the licensee does not exploit, a post-term use ban that prevents the licensee from working an expired patent, the requirement of royalties substantially beyond the life of the patent, or the secrecy of the know-how.
The remaining provisions are not identical to the earlier reg.

goods or services or from goods or services considered by users as interchangeable or substitutable in view of their characteristics, price and intended use, which may in particular occur where the licensee's market share exceeds 40%;

The words, 'the licensee's market share' are defined in article 10:

(9) 'the licensee's market share' means the proportion which the licensed products and other goods or services provided by the licensee, which are considered by users to be interchangeable or substitutable for the licensed products in view of their characteristics, price and intended use, represent the entire market for the licensed products and all other interchangeable or substitutable goods and services in the common market or a substantial part of it;

In most of the group exemptions power was taken to withdraw its benefit in the absence of inter-brand competition.[13] None of the regulations for technology transfer mentions potential competition on the supply side, yet if barriers to entry were low in general, or for particular potential entrants, the Commission might not wish to intervene. The narrow definition does not matter under article 7 in the way it did when the draft technology transfer regulation did not apply if the market share of the licensee exceeded 40 per cent, provided the Commission is willing in practice to take other matters into consideration.

In another way, however, the provision in article 7(1) may be more severe than under the earlier drafts. In those drafts, there was a recital making it clear that the licensee's market share was to be assessed at the date of the licence. It was only when the licensee already enjoyed a 40 per cent share of the market for substitute products that the regulation would not apply. Recital 26 to the actual regulation is less clear:

(26) Where agreements exempted under this Regulation nevertheless have effects incompatible with Article 85(3), the Commission may withdraw the block exemption, in particular where the licensed products are not faced with real competition in the licensed territory (Article 7). This could also be the case where the licensee has a strong position on the market. In assessing the competition, the Commission will pay special attention to cases where the licensee has more than 40% of the whole market for the licensed products and of all the products or services which customers consider interchangeable or substitutable on account of their characteristics, prices and intended use.

---

[13] See, e.g., reg. 2349/84, exempting patent licensing agreements, art. 9(2); reg. 418/85, exempting r & d agreements, art. 10(d); reg. 4087/88, exempting franchising agreements, art. 8(b); reg. 417/85, exempting specialization agreements and reg. 418/85, which exempts collaboration in r & d impose limits of market share which have a similar effect, but are automatic in their operation.

Unlike article 10(17), set out at 9.1.3.4 below, there is no mention of the time at which one should assess the licensee's market share: the date of the licence or when the Commission is deciding whether to withdraw the exemption.

It is hoped that the exemption would not be withdrawn when the licensee was not competing with the licensor at the time of the licence, but later wins a high market share through the use of the technology. This is not so much a question of law, since the list is not exhaustive, but one of the Commission's policy.

I would be concerned about the basic concept of this provision if the Commission were to withdraw the exemption when the licensee achieves a high market share through use of the licensed technology. If the licensor has developed so good an idea that his licensees face no effective competition, should he not be encouraged and rewarded by being allowed to exploit the advantage as he thinks best?

Concern over subsequent market share would be based on static considerations; its underlying thought would be that, once the licensor has innovated and the licensee invested in production facilities and developing a market, it would be more competitive to have more firms in the market. Economists and businessmen, however, think *ex ante*. There must be incentives provided for the licensor to create the technology and for the licensees to invest.[14] Either party might not be prepared to commit its capital and promise to make lump-sum payments or pay royalties unless protected from competition by other members of the network. To withdraw the benefit of the exemption from the big winners must reduce these incentives all round.

An innovator is under commercial pressure to think carefully before giving an exclusive territory to each licensee if it will not be exposed to effective inter-brand competition. If the licensor grants protection it must be because it considers it necessary to induce investment. Since the operation of article 7 is not automatic, these considerations may be argued should the Commission consider intervening.

Where, however, there are two sources of technology, and the holder of one obtains exclusive access to the other or others, then there is much to be said for the Commission intervening, as it did in *Tetra Pak I*.[15]

9.1.3.2 Licensee Refuses Passive Sales (Article 7(2))

The Commission may withdraw the benefit of the exemption where:

> (2) without prejudice to Article 1(1)(6), the licensee refuses, without any

---

[14] See 1.3.1 above.   [15] See n. 1 above.

objectively justified reason, to meet unsolicited orders from users or resellers in the territory of other licensees;

This provision supports the provisions in article 3(3) and (7) which are designed to prevent the discouragement of parallel imports. By virtues of article 1(1)(6) and article 1(2) to (4) the licensee may be restrained from passive sales outside its territory for five years from the first sale within the common market by a licensee. This is circumscribed by article 3(3) when the parties agree to require each other not to sell within their territory to those likely to export to other territories, and by article 3(7) when the territorial restraints permitted by article 1(1) are to last longer than permitted by article 1(2) to (4).

The exemption may be withdrawn if a licensee unilaterally refuses to accept unsolicited orders without justification, even though he is not constrained by contract or concerted practice. Often there may be a commercial justification such as the purchaser's lack of a reputation for creditworthiness in the licensee's territory, or it may not be known that it treats branded products properly.

More difficulty may arise if licensee A refrains from passive sales in the territory of licensee B for fear that B might then suffer overcapacity and dump its products in A's territory. Where markets are concentrated and have been divided historically, oligopolistic inter-dependence may prevent their integration for decades, even in the absence of collusion. Each licensee may act as if he had colluded without it being necessary to do so.

I would be concerned if the Commission were to withdraw its group exemption in such a case. In those circumstances, a refusal to supply may be no more than the individual pursuit of profits. Withdrawal of the exemption would be an inappropriate remedy for a structural problem. Such conduct would be commercially sensible only in concentrated markets. The dissemination of technology helps to break into such markets and should not be discouraged. A firm may be prepared to tool up only because it can rely on such interdependence to appropriate the benefits of its investment.

Where the licensee was already competing with the licensor and had a market share of 40 per cent at the date the licence was granted, the benefit of the regulation can be withdrawn under article 7(1). Article 7(2) is needed only where the market became concentrated because the licensed technology was commercially successful and ousted most competing products or created a new market.

As in article 3(3)(a), the French text in which the regulation was drafted refers to supplying customers 'established' in the territory of other licensees. The word 'established' was not inserted in the English text either in the know-how or the technology transfer regulation. Both texts would be relevant in construing the regulation before the Community Court.

9.1.3.3 Impeding Cross-frontier Trade (Article 7(3))

The next instance is also derived from the earlier regulations. It reinforces article 3(3). To fall foul of the black list, there must be collusion to deter trade across boundaries, but article 7(3) enables the Commission to withdraw the exemption where there is unilateral conduct by each party to the licence, even if not collusive. The Commission may withdraw the exemption where:

> (3) the parties:
>> (a) without any objectively justified reason, refuse to meet orders from users or resellers[16] in their respective territories who would market the products in other territories within the common market; or
>> (b) make it difficult for users or resellers to obtain the products from other resellers within the common market, and in particular where they exercise intellectual property rights or take measures so as to prevent resellers or users from obtaining outside, or from putting on the market in the licensed territory products which have been lawfully put on the market within the common market by the licensor or with his consent;

The question whether such conduct by one party only enables the Commission to withdraw the exemption is not important, since it has power to do so under the introductory words of article 7.

9.1.3.4 Where the Agreement is Horizontal and Forecloses Other Technologies (Article 7(4))

The final example specified is where:

> (4) the parties were competing manufacturers at the date of the grant of the licence and obligations on the licensee to produce a minimum quantity or to use his best endeavours as referred to in Article 2(1), (9) and (17) respectively have the effect of preventing the licensee from using competing technologies.

Unlike article 7(1) it is made clear that the time to determine whether the parties were competing manufacturers is the date the licence was granted.

At 1.4.2, 5.1.1, 6.2, and elsewhere, I questioned whether horizontal agreements should have been excluded from the regulation. Although article 7(4) applies only to agreements between those who were already competing when the licence was granted, it is arguable that under the

---

[16] Again, the French text inserts the word 'established' at this point.

general words the Commission might withdraw the exemption, where they were strong potential competitors.[17]

I have stated more than once[18] that where the parties were competitors at the date of the licence, the promise by the licensee to use its best endeavours, to pay a minimum royalty, or exploit the technology beyond a minimum may, indeed, enable the parties to divide the common market between them. The no-competition item in the black list is the only one that applies to competing technology or products made thereby and these provisions limit the effects of that provision. If the licensee is discouraged thereby from exploiting other technology, and the licensor lacks other technology, such clauses enable them to agree to keep out of each other's territory for ten years from the date of the first licence in the licensee's territory.

The term 'competing manufacturers' is defined in article 10(17):

> (17) 'competing manufacturers' or manufacturers of 'competing products' means manufacturers who sell products which, in view of their characteristics, price and intended use, are considered by users to be interchangeable or substitutable for the licensed products.

Again, the failure to include the supply side of the market does not matter in the context of article 7, since the parties have an opportunity to express their views to the Commission before it makes a formal decision.

9.1.4 Conclusion on the Withdrawal of the Exemption

The concept of withdrawing a group exemption is sensible. In a group exemption, officials do not know all the circumstances relevant to the grant of an individual exemption. Clauses that would not be anti-competitive, and indeed might well be pro-competitive, in a vertical agreement may be anti-competitive where the parties are already competitors or potential competitors, and the licensee has significant market power short of a dominant position. For this reason I approve of the basic idea of article 7(4). Most of the other specific examples, however, seem to me unfortunate.

---

[17] In *Re the Agreements between BBC Brown Boveri and NGK Insulators Ltd* (88/54/EEC), [1988] OJ L301/68, [1989] 4 CMLR 610, CMR 11,305, the Commission found that a party to a joint venture which was in a quite different industry became a potential competitor of the other party because of the licence the latter granted it. Consequently, the licence to the Japanese parent to manufacture and sell only in the Far East restricted competition contrary to art. 85(1) and had to be exempted. It is thought that the decision under art. 85(1) was wrong. In any event, it should not be applied under art. 7(8), which refers to being competitors before the licence was granted. The Commission is required to look *ex ante*.

[18] e.g., 5.1.1 and 7.3.9.3 above.

The lack of competing technology listed in article 7(1) makes any territorial or other protection important, but, unless the parties were competitors when the licence was granted, or have bought up competing technologies, the agreement is unlikely to have reduced competition. Indeed, by providing for exploitation it has increased it. The possibility of withdrawing exemptions on this ground may make licensing more risky, since it is only where innovations are highly successful that the exemption is likely to be withdrawn.

Paragraphs (2) and (3) may be difficult to apply in practice. The Commission is concerned to persuade firms to treat the common market as integrated, but if unilateral action causes the Commission to withdraw the exemption, it may become harder for firms to operate rationally in concentrated markets. If the Commission cannot think of better examples when preparing exemptions, it might be better to keep just the general words.

9.2 TRANSITIONAL PROVISIONS

The transitional provisions are much simpler than under the earlier regulations. There can be few licences that qualified under their terms that do not qualify under regulation 240/96.[19] The territorial protection permitted under the new regulation cannot be shorter; the white list is more extensive, the black list has been shortened, and some items moved from there to the opposition procedure. Consequently, there was a possibility that nothing would be done to preserve the legality of agreements that qualified under the earlier regulations when the new one came into force on 1 April 1996.

In fact, however, licences that were exempt under the earlier regulations have been preserved under the new regulation until its expiry at the end of March 2006. Consequently, there is no need for lawyers to scrutinize all their licensing agreements to check that they still qualify.

9.2.1 Replacement of the Earlier Block Exemptions

Recital 3 states that the two earlier group exemptions should be combined into a single regulation and the rules affecting know-how and patent licensing harmonized and simplified as far as possible in order to encourage the dissemination of technical knowledge in the Community and promote the manufacture of more sophisticated products.

---

[19] One example is *Delta Chemie* which may not apply to technology transfer. Another is that it is not only the territorial protection that cannot be given against a sales licensee or franchisee, but the reg. as a whole.
These are not likely to be very important in practice.

The know-how regulation was due to expire only at the end of the millennium, but has been repealed from 1 April 1996 by article 11. The patent regulation expired at the end of 1994, and was retrospectively prolonged until the end of June by regulation 70/95 of 17 January 1995.[20] It was further prolonged retrospectively on 7 September until the end of the year by regulation 2131/95.[21] The Commission's progress with the new regulation was further delayed, and the patent regulation had to be renewed yet again by article 11(2) of the technology transfer regulation.

Normally courts are hostile towards retrospective legislation, but where an exemption is prolonged because the regulation to replace it is not ready it would probably be upheld, on the ground of certainty of law or legitimate expectations. The question might arise if proceedings were started in a national court during one of the periods when there was no patent regulation in existence.[22]

Article 11 provides:

1. Regulation (EEC) No 556/89 is hereby repealed with effect from 1 April 1996.
2. Regulation (EEC) No 2349/84 shall continue to apply until 31 March 1996.
3. The prohibition in Article 85(1) of the Treaty shall not apply to agreements in force on 31 March 1996 which fulfil the exemption requirements laid down by Regulation (EEC) No 2349/84 or (EEC) No 556/89.[23]

Article 13 provides for article 11(2) to come into force at the beginning of 1996 so as to ensure that the patent regulation is renewed until the end

---

[20] [1995] OJ L12/13. This was implemented for the EEA by its Joint Committee in decision 23/95, amending Annex XIV (Competition) to the EEA Agreement [1995] OJ L.139/14. The EFTA Member States will have to incorporate the reg. into their national laws.

[21] [1995] OJ L214/6. This was implemented for the EEA by its Joint Committee in decision 65/95, amending Annex XIV (Competition) to the EEA Agreement [1996] OJ L8/36.

[22] **Chris Kerse** said at a conference on the reg. organized by IBC on 22 May 1996, since published as 'Block Exemptions Under Article 85(3): The Technology Transfer Regulation—Procedural Issues' [1996] ECLR 331:

'Retroactivity is not always prohibited under Community law. The Court of Justice has taken the position that legal certainty is a force against retroactivity of Community legislation but has accepted retroactivity where there are compelling reasons for it. Normally provisions granting rights to individuals cannot be retroactively withdrawn and the practice of the Commission as regards block exemptions has been to provide safeguards for parties to agreements in the form of transitional relief . . . Such relief is not, it will be appreciated, retroactive. On the other hand the "revival" and continuation of Regulation 2349/84 clearly is, and although it has not been challenged and might indeed survive challenge it is not a good precedent for Community legislation to follow. Parties to relevant agreements should not have had to suffer unnecessary legal uncertainty and the Commission should have taken the necessary legal measures promptly, notwithstanding any potential political embarrassment.'

[23] This leaves no gap: see art. 4(3) of reg. 1182/71, determining the rules applicable to periods, dates, and time limits.

of March 1996 and that agreements exempt under its terms remain exempt by virtue of article 11(3).

If a licence did not qualify under the patent or know-how regulations, but does qualify under regulation 240/96, the provisions for exclusivity and the associated export bans, and other provisions that were exempted or came within the former black lists may suddenly have become retrospectively valid from 1 April 1996. This is an odd concept, but no odder than the similar result that applies when a new agreement is granted an individual exemption.

9.2.2 Reassessment of the Technology Transfer Regulation (Article 12)

As mentioned at 1.4.2.1, the Commission is currently preparing a draft Green Book on vertical restraints in distribution agreements of all kinds—exclusive, selective, and franchising. The original plan was to set out various options that were possible for the future, ranging from continuation of the present practice of treating any restriction on a firm's freedom that has appreciable effects as contrary to article 85(1) to making a full market analysis before deciding that an agreement has even the object of restricting competition.

If the final option were chosen in the White Book which it is hoped will follow the Green Book, the need for group exemptions might decrease. As this Chapter is being written, the Green Book has not been published and it is unlikely that a very liberal option will be selected, unless industry is successful in lobbying. Whether or not in expectation of a possible change in its practice, the Commission has committed itself to reassessing the technology transfer regulation before its expiry in 2006. Article 12 provides:

> 1. The Commission shall undertake regular assessments of the application of this Regulation, and in particular of the opposition procedure provided for in Article 4.
> 2. The Commission shall draw up a report on the operation of this Regulation before the end of the fourth year following its entry into force and shall, on that basis, assess whether any adaptation of the Regulation is desirable.

9.2.3 Entry into Force (Article 13)

The regulation is to remain in force for ten years starting on 1 April 1996. Article 13 provides:

> This Regulation shall enter into force on 1 April 1996.
> It shall apply until 31 March 2006.

Article 11(2) of this Regulation shall, however, enter into force on 1 January 1996.

This was necessary to prolong the operation of the patent licensing regulation.

9.2.4 Conclusion

The transitional provisions are very simple in the technology transfer regulation because its terms are so much more liberal, that there was no need to do more than preserve the legality of agreements that qualified under the earlier regulations to save firms the expense of vetting all their earlier licences to ensure that they qualify under the new law.

10
Conclusion

There is little to be said in conclusion. The technology transfer regulation is welcome. It applies to more agreements than did the earlier regulations, and is drafted more simply. It applies to pure patent and pure know-how agreements as well as mixed licences, and there is no need, as there was under the patent regulation, to consider whether the know-how is ancillary to the patented technology. It will be easier and faster to understand the new regulation.

The list of intellectual property rights that are to be treated as patents has been lengthened by the addition of plant breeders' rights and supplementary protection certificates for medicinal products. The definition of 'ancillary provisions' has been drafted in such a way that industrial franchising is probably not excluded from the regulation when sufficient know-how or patents are licensed.

When a similar decision is adopted by the Joint Committee of the EEA, probably in the summer or autumn of 1996, most agreements will come within that rather than the regulation I have analysed, but the wording in English is likely to be the same.

The periods of territorial protection permitted in pure and mixed know-how licences all start from the same time. The date when the licensed products were first put on the market in the common market by a licensee cannot be earlier than the first licence or the date when the products were first put on the market by licensor or a licensee within the common market and may well be considerably later. So the territorial protection permitted cannot be shorter and may be much longer than under the earlier regulations.

Nevertheless, the focus is wrong. The regulation defines the restrictions that may be imposed on a licensee's conduct rather than the licensees who may be protected. So, the protection against passive sales may inure for the benefit of a distributor against a manufacturing licensee but not against a distributor to protect a licensee who produces the licensed products. Moreover, the inability to control parallel trade either by the use of intellectual property rights or by contractual provisions works in a discriminatory way. There is considerable protection when the products are tailor-made for the client or where, for other reasons, they do not pass through dealers but are delivered direct to customers. There may also be considerable protection for products where freight forms a large part of the delivered value, but not for objects that are smaller or more valuable. This is one of the drawbacks of the Commission's bifurcation of article 85(1) and (3), resulting in the need for formalistically drafted group exemptions.

Several additions to the white list in article 2 are important. One of the most interesting is article 2(1)(14) which implies that the grant of a patent licence does not exhaust the right. This may well apply also to other intellectual property rights that protect investment in innovation such as copyright or design rights. It may not apply to trade marks which protect reputation.

The black list has been reduced from twelve items in the know-how regulation to seven. Where the licensee could not have entered the market without a licence from the technology holder, I see nothing wrong with restrictions on price, quantity, or customers. All the reasons that have caused field-of-use provisions to be whitelisted apply to such restrictions that affect only the licensed technology, but this has not been accepted by the Commission.

Two provisions that used to be blacklisted have been made subject to the opposition procedure. The Commission no longer objects to tying for the purpose of monitoring royalties, and quality specifications that may be stricter than those required of other licensees. This may be useful when higher quality is required by some customers than by others. The Commission's attitude to no-challenge clauses is also becoming more relaxed. There are many provisions where the Commission now perceives *ex ante* rather than *ex post*. Licensors without sufficient production capacity to supply the whole common market may be deterred from licensing their technology, if this might open them to expensive patent litigation. The opposition procedure has the great advantage of enabling the Commission to give validity to agreements without having to make an individual decision. Under this regulation, the Commission has promised to waive the need to provide more market analysis than is directly available, so notification should be much easier and cheaper.

Signor Guttuso, who used to devote his time to handling technology transfer agreements has taken on agreements in the media sector. I infer that the Commission does not expect there to be much work on licensing agreements and think that the regulation may be applied liberally.

If the White Book on vertical restraints should opt for a policy of treating as forbidden by article 85(1) only those agreements that have the object or effect of significantly restricting competition perceived in the context of a market analysis: if vertical agreements escape the prohibition in the absence of market power at either level of trade and industry, the regulation may become largely redundant. It is, however, possible that this may not happen for a long time, and meanwhile the group exemption provides a safe harbour for many technology transfer agreements.

Appendix I

Excerpts from the Treaty establishing the European Community[1]

Part One—Principles

Article 2

The Community shall have as its task, by establishing a common market and *an economic and monetary union and by implementing the common policies or activities referred to in articles 3 and 3a*, to promote throughout the Community a harmonious *and balanced* development of economic activities, *sustainable and non-inflationary growth respecting the environment, a high degree of convergence of economic employment, a high level of employment and of social protection, the* raising of the standard of living and *quality of life, and economic and social cohesion and solidarity among Member States.*

Article 3

For the purposes set out in Article 2, the activities of the Community shall include, as provided in this Treaty and in accordance with the timetable set out therein:

(a) the elimination, as between Member States, of customs duties and of quantitative restrictions on the import and export of goods, and of all other measures having equivalent effect;

(b) *a common commercial policy;*

(c) *an internal market characterised by the abolition, as between Member States, of obstacles to the free movement of goods, persons, services and capital;*

(d) *measures concerning the entry and movement of persons in the internal market as provided for in Article 100(c);*

(e) a common policy in the sphere of agriculture *and fisheries;*

(f) the adoption of a common policy in the sphere of transport;

(g) the institution of a system ensuring that competition in the *internal* market is not distorted;

(g) the application of procedures by which the economic policies of Member States can be coordinated and disequilibria in their balances of payments remedied;

(h) the approximation of the laws of Member States to the extent required for the proper functioning of the common market;

[1] The additions made by the Maastricht Treaty have been inserted in italics and the deletions made.

(i) the creation of a European Social Fund in order to improve employment opportunities for workers and to contribute to the raising of their standard of living;

(j) the establishment of a European Investment Bank to facilitate the economic expansion of the Community by opening up fresh resources;

(k) the association of the overseas countries and territories in order to increase trade and to promote jointly economic and social development.

Article 5

Member States shall take all appropriate measures, whether general or particular, to ensure fulfilment of the obligations arising out of this Treaty or resulting from action taken by the institutions of the Community. They shall facilitate the achievement of the Community's tasks.

They shall abstain from any measure which could jeopardise the attainment of the objectives of this Treaty.

CHAPTER 2—ELIMINATION OF QUANTITATIVE RESTRICTIONS BETWEEN MEMBER STATES

Article 30

Quantitative restrictions on imports and all measures having equivalent effect shall, without prejudice to the following provisions, be prohibited between Member States.

Article 36

The provisions of Articles 30–36 shall not preclude prohibitions or restrictions on imports, exports or goods in transit justified on grounds of public morality, public policy or public security; the protection of health and life of humans, animals or plants; the protection of national treasures possessing artistic, historic or archæological value; or the protection of industrial and commercial property. Such prohibitions or restrictions shall not, however, constitute a means of arbitrary discrimination or a disguised restriction on trade between Member States.

Part Three—Policy of the Community

TITLE I—COMMON RULES
CHAPTER 1—RULES ON COMPETITION
SECTION 1—RULES APPLYING TO UNDERTAKINGS

Article 85

1. The following shall be prohibited as incompatible with the common market: all agreements between undertakings, decisions by associations of undertakings

and concerted practices which may affect trade between Member States and which have as their object or effect the prevention, restriction or distortion of competition within the common market, and in particular those which:

(a) directly or indirectly fix purchase or selling prices or any other trading conditions;

(b) limit or control production, markets, technical development, or investment;

(c) share markets or sources of supply;

(d) apply dissimilar conditions to equivalent transactions with other trading parties, thereby placing them at a competitive disadvantage;

(e) make the conclusion of contracts subject to acceptance by the other parties of supplementary obligations which, by their nature or according to commercial usage, have no connection with the subject of such contracts.

2. Any agreements or decisions prohibited pursuant to this Article shall be automatically void.

3. The provisions of paragraph 1 may, however, be declared inapplicable in the case of:

— any agreement or category of agreements between undertakings;

— any decision or category of decisions by associations of undertakings;

— any concerted practice or category of concerted practices;

which contributes to improving the production or distribution of goods or to promoting technical or economic progress, while allowing consumers a fair share of the resulting benefit, and which does not:

(a) impose on the undertakings concerned restrictions which are not indispensable to the attainment of these objectives;

(b) afford such undertakings the possibility of eliminating competition in respect of a substantial part of the products in question.

Article 86

Any abuse by one or more undertakings of a dominant position within the common market or in a substantial part of it shall be prohibited as incompatible with the common market in so far as it may affect trade between Member States. Such abuse may, in particular, consist in:

(a) directly or indirectly imposing unfair purchase or selling prices or other unfair trading conditions;

(b) limiting production, markets or technical development to the prejudice of consumers;

(c) applying dissimilar conditions to equivalent transactions with other trading parties, thereby placing them at a competitive disadvantage;

(d) making the conclusion of contracts subject to acceptance by the other parties of supplementary obligations which, by their nature or according to commercial usage, have no connection with the subject of such contracts.

Part Six—General and Final Provisions

Article 222

This Treaty shall in no way prejudice the rules of Member States governing the system of property ownership.

Excerpts from the Treaty establishing the European Economic Area

Article 53

1. The following shall be prohibited as incompatible with the functioning of this Agreement: all agreements between undertakings, decisions by associations of undertakings and concerted practices which may affect trade between Contracting Parties and which have as their object or effect the prevention, restriction or distortion of competition within the territory covered by this Agreement, and in particular those which:

(a) directly or indirectly fix purchase or selling prices or any other trading conditions;

(b) limit or control production, markets, technical development, or investment;

(c) share markets or sources of supply;

(d) apply dissimilar conditions to equivalent transactions with other trading parties, thereby placing them at a competitive disadvantage;

(e) make the conclusion of contracts subject to acceptance by the other parties of supplementary obligations which, by their nature or according to commercial usage, have no connection with the subject of such contracts.

2. Any agreements or decisions prohibited pursuant to this Article shall be automatically void.

3. The provisions of paragraph 1 may, however, be declared inapplicable in the case of:

— any agreement or category of agreements between undertakings;
— any decision or category of decisions by associations of undertakings;
— any concerted practice or category of concerted practices;

which contributes to improving the production or distribution of goods or to promoting technical or economic progress, while allowing consumers a fair share of the resulting benefit, and which does not:

(a) impose on the undertakings concerned restrictions which are not indispensable to the attainment of these objectives;

(b) afford such undertakings the possibility of eliminating competition in respect of a substantial part of the products in question.

Article 54

Any abuse by one or more undertakings of a dominant position within the territory covered by this Agreement or in a substantial part of it shall be prohibited as incompatible with the functioning of this Agreement in so far as it may affect trade between Contracting Parties. Such abuse may, in particular, consist in:

(a) directly or indirectly imposing unfair purchase or selling prices or other unfair trading conditions;

(b) limiting production, markets or technical development to the prejudice of consumers;

(c) applying dissimilar conditions to equivalent transactions with other trading parties, thereby placing them at a competitive disadvantage;

(d) making the conclusion of contracts subject to acceptance by the other parties of supplementary obligations which, by their nature or according to commercial usage, have no connection with the subject of such contracts.

Protocol 28 on Intellectual Property

Article 2. Exhaustion of rights

1. To the extent that exhaustion is dealt with in Community measures or jurisprudence, the Contracting Parties shall provide for such exhaustion of intellectual property rights as is laid down in Community law. Without prejudice to future developments of case-law, this provision shall be interpreted in accordance with the meaning established in the relevant rulings of the Court of Justice of the European Commuities given prior to the signature of this agreement.

2. As regards patent rights, this provision shall take effect at the latest one year after the entry into force of this Agreement.

Appendix II—Commission Regulation (EC) No 240/96* of 31 January 1996 on the application of Article 85(3) of the Treaty to certain categories of technology transfer agreements (Text with EEA relevance)

THE COMMISSION OF THE EUROPEAN COMMUNITIES

Having regard to the Treaty establishing the European Community,

Having regard to Council Regulation No 19/65/EEC of 2 March 1965 on the application of Article 85(3) of the Treaty to certain categories of agreements and concerted practices,[1] as last amended by the Act of Accession of Austria, Finland and Sweden, and in particular Article 1 thereof,

Having published a draft of this Regulation,[2]

After consulting the Advisory Committee on Restrictive Practices and Dominant Positions,

Whereas:

Vires

(1) Regulation No 19/65/EEC empowers the Commission to apply Article 85(3) of the Treaty by regulation to certain categories of agreements and concerted practices falling within the scope of Article 85(1) which include restrictions imposed in relation to the acquisition or use of industrial property rights—in particular of patents, utility models, designs or trademarks—or to the rights arising out of contracts for assignment of, or the right to use, a method of manufacture of knowledge relating to use or to the application of industrial processes.

---

* The annotations in bold italics and italics are the author's. The references to articles in the recitals are to those mainly affected by each recital. Those in the articles are to the governing recitals. The references at the end of provisions are to other provisions which must be considered before advice can be given.

[1] [1965] JO 533.　　　　[2] [1994] OJ C178/3.

Former regulations

(2) The Commission has made use of this power by adopting Regulation (EEC) No 2349/84 of 23 July 1984 on the application of Article 85(3) of the Treaty to certain categories of patent licensing agreements,[3] as last amended by Regulation (EC) No 2131/95,[4] and Regulation (EEC) No 556/89 of 30 November 1988 on the application of Article 85(3) of the Treaty to certain categories of know-how licensing agreements,[5] as last amended by the Act of Accession of Austria, Finland and Sweden.

Unify the two regulations and repeal them (Article 11)

(3) These two block exemptions ought to be combined into a single regulation covering technology transfer agreements, and the rules governing patent licensing agreements and agreements for the licensing of know-how ought to be harmonized and simplified as far as possible, in order to encourage the dissemination of technical knowledge in the Community and to promote the manufacture of technically more sophisticated products. In those circumstances, Regulation (EEC) No 556/89 should be repealed.

Patents, know-how and mixed licences—definitions (Articles 1(1) and 10)

(4) This Regulation should apply to the licensing of Member States' own patents, Community patents[6] and European patents[7] ('pure' patent licensing agreements). It should also apply to agreements for the licensing of non-patented technical information such as descriptions of manufacturing processes, recipes, formulae, designs or drawings, commonly termed 'know-how' ('pure' know-how licensing agreements), and to combined patent and know-how licensing agreements ('mixed' agreements), which are playing an increasingly important role in the transfer of technology. For the purposes of this Regulation, a number of terms are defined in Article 10.

Patent and know-how licences qualifying for group exemption defined

(5) Patent or know-how licensing agreements are agreements whereby one undertaking which holds a patent or know-how ('the licensor') permits another undertaking ('the licensee') to exploit the patent thereby licensed, or communicates the know-how to it, in particular for purposes of manufacture, use or putting on the market. In the light of experience acquired so far, it is possible to define a category of licensing agreements covering all or part of the common market which are capable of falling within the scope of Article 85(1) but which can normally be regarded as satisfying the conditions laid down in Article 85(3), where patents are necessary for the achievement of the objects of the licensed technology by a mixed agreement or where know-how—whether it is ancillary to patents or independent of them—is secret, substantial and identified in any appropriate form. These criteria are intended only to ensure that the licensing of the know-how or the grant

---

[3] [1984] OJ L219/15.   [4] [1995] OJ L214/6.   [5] [1989] OJ L61/1.
[6] Community Patent Convention [1976] OJ L17/1.
[7] European Patent Convention of 5 Oct. 1993.

of the patent licence justifies a block exemption of obligations restricting competition. This is without prejudice to the right of the parties to include in the contract provisions regarding other obligations, such as the obligation to pay royalties, even if the block exemption no longer applies. *(Article 10(1)–(6) and (15))*

Other ancillary intellectual property rights included (Article 5(1)(4))

(6) It is appropriate to extend the scope of this Regulation to pure or mixed agreements containing the licensing of intellectual property rights other than patents (in particular, trademarks, design rights and copyright, especially software protection), when such additional licensing contributes to the achievement of the objects of the licensed technology and contains only ancillary provisions. *(Article 10(15))*

Extraterritorial application

(7) Where such pure or mixed licensing agreements contain not only obligations relating to territories within the common market but also obligations relating to non-member countries, the presence of the latter does not prevent this Regulation from applying to the obligations relating to territories within the common market. Where licensing agreements for non-member countries or for territories which extend beyond the frontiers of the Community have effects within the common market which may fall within the scope of Article 85(1), such agreements should be covered by this Regulation to the same extent as would agreements for territories within the common market.

Excludes sales licences, franchises, patent pools and so on (Article 5(1))

(8) The objective being to facilitate the dissemination of technology and the improvement of manufacturing processes, this Regulation should apply only where the licensee himself manufactures the licensed products or has them manufactured for his account, or where the licensed product is a service, provides the service himself or has the service provided for his account, irrespective of whether or not the licensee is also entitled to use confidential information provided by the licensor for the promotion and sale of the licensed product. The scope of this Regulation should therefore exclude agreements solely for the purpose of sale. Also to be excluded from the scope of this Regulation are agreements relating to marketing know-how communicated in the context of franchising arrangements and certain licensing agreements entered into in connection with arrangements such as joint ventures or patent pools and other arrangements in which a licence is granted in exchange for other licences not related to improvements to or new applications of the licensed technology. Such agreements pose different problems which cannot at present be dealt with in a single regulation (Article 5)

Assignments, sub-licences (Article 6)

(9) Given the similarity between sale and exclusive licensing, and the danger that the requirements of this Regulation might be evaded by presenting as assignments what are in fact exclusive licences restrictive of competition, this Regulation should apply to agreements concerning the assignment and acquisition

of patents or know-how where the risk associated with exploitation remains with the assignor. It should also apply to licensing agreements in which the licensor is not the holder of the patent or know-how but is authorized by the holder to grant the licence (as in the case of sub-licences) and to licensing agreements in which the parties' rights or obligations are assumed by connected undertakings (Article 6).

Open exclusive licences

(10) Exclusive licensing agreements, i.e. agreements in which the licensor undertakes not to exploit the licensed technology in the licensed territory himself or to grant further licences there, may not be in themselves incompatible with Article 85(1) where they are concerned with the introduction and protection of a new technology in the licensed territory, by reason of the scale of the research which has been undertaken, of the increase in the level of competition, in particular inter-brand competition, and of the competitiveness of the undertakings concerned resulting from the dissemination of innovation within the Community. In so far as agreements of this kind fall, in other circumstances, within the scope of Article 85(1), it is appropriate to include them in Article 1 in order that they may also benefit from the exemption. *(Nungesser, points 53–67)*

Export bans and exhaustion (Articles 1(1) and 2(1)(14))

(11) The exemption of export bans on the licensor and on the licensees does not prejudice any developments in the case law of the Court of Justice in relation to such agreements, notably with respect to Articles 30 to 36 and Article 85(1). This is also the case, in particular, regarding the prohibition on the licensee from selling the licensed product in territories granted to other licensees (passive competition).

Benefits of territorial restraints—duration of pure patent licences (Article 1(2))

(12) The obligations listed in Article 1 generally contribute to improving the production of goods and to promoting technical progress. They make the holders of patents or know-how more willing to grant licences and licensees more inclined to undertake the investment required to manufacture, use and put on the market a new product or to use a new process. Such obligations may be permitted under this Regulation in respect of territories where the licensed product is protected by patents as long as these remain in force.

Duration of territorial restraints in pure know-how licences (Article 1(3))

(13) Since the point at which the know-how ceases to be secret can be difficult to determine, it is appropriate, in respect of territories where the licensed technology comprises know-how only, to limit such obligations to a fixed number of years. Moreover, in order to provide sufficient periods of protection, it is appropriate to take as the starting-point for such periods the date on which the product is first put on the market in the Community by a licensee.

Longer territorial restraints—individual exemptions

(14) Exemption under Article 85(3) of longer periods of territorial protection for know-how agreements, in particular in order to protect expensive and risky investment or where the parties were not competitors at the date of the grant of the licence, can be granted only by individual decision. On the other hand, parties are

free to extend the term of their agreements in order to exploit any subsequent improvement and to provide for the payment of additional royalties. However, in such cases, further periods of territorial protection may be allowed only starting from the date of licensing of the secret improvements in the Community, and by individual decision. Where the research for improvements results in innovations which are distinct from the licensed technology the parties may conclude a new agreement benefiting from an exemption under this Regulation. *(Articles 3(7) and 8(3))*

Passive sales (Article 1(1)(5) and (6))

(15) Provision should also be made for exemption of an obligation on the licensee not to put the product on the market in the territories of other licensees, the permitted period for such an obligation (this obligation would ban not just active competition but passive competition too) should, however, be limited to a few years from the date on which the licensed product is first put on the market in the Community by a licensee, irrespective of whether the licensed technology comprises know-how, patents or both in the territories concerned. *(Articles 1(2)–(4) and 2(1)(14))*

Duration of territorial restraints in mixed licences (Article 1(4))

(16) The exemption of territorial protection should apply for the whole duration of the periods thus permitted, as long as the patents remain in force or the know-how remains secret and substantial. The parties to a mixed patent and know-how licensing agreement must be able to take advantage in a particular territory of the period of protection conferred by a patent or by the know-how, whichever is the longer.

Excessive territorial restraints (Article 3(3))

(17) The obligations listed in Article 1 also generally fulfil the other conditions for the application of Article 85(3). Consumers will, as a rule, be allowed a fair share of the benefit resulting from the improvement in the supply of goods on the market. To safeguard this effect, however, it is right to exclude from the application of Article 1 cases where the parties agree to refuse to meet demand from users or resellers within their respective territories who would resell for export, or to take other steps to impede parallel imports. The obligations referred to above thus only impose restrictions which are indispensable to the attainment of their objectives. *(Article 7(3))*

White list (Article 2)

(18) It is desirable to list in this Regulation a number of obligations that are commonly found in licensing agreements but are normally not restrictive of competition, and to provide that in the event that because of the particular economic or legal circumstances they should fall within Article 85(1), they too will be covered by the exemption. This list, in Article 2, is not exhaustive.

Black list (Article 3)

(19) This Regulation must also specify what restrictions or provisions may not be included in licensing agreements if these are to benefit from the block exemption.

The restrictions listed in Article 3 may fall under the prohibition of Article 85(1), but in their case there can be no general presumption that, although they relate to the transfer of technology, they will lead to the positive effects required by Article 85(3), as would be necessary for the granting of a block exemption. Such restrictions can be declared exempt only by an individual decision, taking account of the market position of the undertakings concerned and the degree of concentration on the relevant market.

Post-term use ban—feed- and grant-back (Articles 2(1)(3) and (4) and 3(6))

(20) The obligations on the licensee to cease using the licensed technology after the termination of the agreement (Article 2(1)(3)) and to make improvements available to the licensor (Article 2(1)(4)) do not generally restrict competition. The post-term use ban may be regarded as a normal feature of licensing, as otherwise the licensor would be forced to transfer his know-how or patents in perpetuity. Undertakings by the licensee to grant back to the licensor a licence for improvements to the licensed know-how and/or patents are generally not restrictive of competition if the licensee is entitled by the contract to share in future experience and inventions made by the licensor. On the other hand, a restrictive effect on competition arises where the agreement obliges the licensee to assign to the licensor rights to improvements of the originally licensed technology that he himself has brought about (Article 3(6)).

Royalties (Articles 2(1)(7) and 3)

(21) The list of clauses which do not prevent exemption also includes an obligation on the licensee to keep paying royalties until the end of the agreement independently of whether or not the licensed know-how has entered into the public domain through the action of third parties or of the licensee himself (Article 2(1)(7)). Moreover, the parties must be free, in order to facilitate payment, to spread the royalty payments for the use of the licensed technology over a period extending beyond the duration of the licensed patents, in particular by setting lower royalty rates. As a rule, parties do not need to be protected against the foreseeable financial consequences of an agreement freely entered into, and they should therefore be free to choose the appropriate means of financing the technology transfer and sharing between them the risks of such use. However, the setting of rates of royalty so as to achieve one of the restrictions listed in Article 3 renders the agreement ineligible for the block exemption.

Field-of-use restriction (Article 2(1)(8))

(22) An obligation on the licensee to resrict his exploitation of the licensed technology to one or more technical fields of application ('fields of use') or to one or more product markets is not caught by Article 85(1) either, since the licensor is entitled to transfer the technology only for a limited purpose (Article 2(1)(8)). *(Article 3(4))*

Customer allocation (Articles 3(4), 4(1) and 2(1)(13))

(23) Clauses whereby the parties allocate customers within the same technological field of use or the same product market, either by an actual prohibition on

supplying certain classes of customer or through an obligation with an equivalent effect, would also render the agreement ineligible for the block exemption where the parties are competitors for the contract products (Article 3(4)). Such restrictions between undertakings which are not competitors remain subject to the opposition procedure. Article 3 does not apply to cases where the patent or know-how licence is granted in order to provide a single customer with a second source of supply. In such a case, a prohibition on the second licensee from supplying persons other than the customer concerned is an essential condition for the grant of a second licence, since the purpose of the transaction is not to create an independent supplier in the market. The same applies to limitations on the quantities the licensee may supply to the customer concerned (Article 2(1)(13))

Price and quantity restrictions (Articles 2(1)(12)–(13) and 3(1), (4), and (5))

(24) Besides the clauses already mentioned, the list of restrictions which render the block exemption inapplicable also includes restrictions regarding the selling prices of the licensed product or the quantities to be manufactured or sold, since they seriously limit the extent to which the licensee can exploit the licensed technology and since quantity restrictions particularly may have the same effect as export bans (Article 3(1) and (5)). This does not apply where a licence is granted for use of the technology in specific production facilities and where both a specific technology is communicated for the setting-up, operation and maintenance of these facilities and the licensee is allowed to increase the capacity of the facilities or to set up further facilities for its own use on normal commercial terms. On the other hand, the licensee may lawfully be prevented from using the transferred technology to set up facilities for third parties, since the purpose of the agreement is not to permit the licensee to give other producers access to the licensor's technology while it remains secret or protected by patent (Article 2(1)(12)).

Opposition procedure (Article 4)

(25) Agreements which are not automatically covered by the exemption because they contain provisions that are not expressly exempted by this Regulation and not expressly excluded from exemption, including those listed in Article 4(2), may, in certain circumstances, nonetheless be presumed to be eligible for application of the block exemption. It will be possible for the Commission rapidly to establish whether this is the case on the basis of the information undertakings are obliged to provide under Commission Regulation (EC) No 3385/94.[8] The Commission may waive the requirement to supply specific information required in form A/B but which it does not deem necessary. The Commission will generally be content with communication of the text of the agreement and with an estimate, based on directly available data, of the market structure and of the licensee's market share. Such agreements should therefore be deemed to be covered by the exemption provided for in this Regulation where they are notified to the Commission and the Commission does not oppose the application of the exemption within a specified period of time. *(Article 9)*

---

[8] [1994] OJ L377/28.

Withdrawal of exemption (Article 7)

(26) Where agreements exempted under this Regulation nevertheless have effects incompatible with Article 85(3), the Commission may withdraw the block exemption, in particular where the licensed products are not faced with real competition in the licensed territory (Article 7). This could also be the case where the licensee has a strong position on the market. In assessing the competition, the Commission will pay special attention to cases where the licensee has more than 40% of the whole market for the licensed products and of all the products or services which customers consider interchangeable or substitutable on account of their characteristics, prices and intended use.

No need to notify

(27) Agreements which come within the terms of Articles 1 and 2 and which have neither the object nor the effect of restricting competition in any other way need no longer be notified. Nevertheless, undertakings will still have the right to apply in individual cases for negative clearance or for exemption under Article 85(3) in accordance with Council Regulation No 17,[9] as last amended by the Act of Accession of Austria, Finland and Sweden. They can in particular notify agreements obliging the licensor not to grant other licences in the territory, where the licensee's market share exceeds or is likely to exceed 40%,

HAS ADOPTED THIS REGULATION:

Article 1

Territorial restrictions (Recitals 3–10 and 12)

1. Pursuant to Article 85(3) of the Treaty and subject to the conditions set out below, it is hereby declared that Article 85(1) of the Treaty shall not apply to pure patent licensing or know-how licensing agreements and to mixed patent and know-how licensing agreements, including those agreements containing ancillary provisions relating to intellectual property rights other than patents, to which only two undertakings are party and which include one or more of the following obligations: *(Articles 1(2)–(5), (5) and 10(1)–(8))*

Sole licence

(1) an obligation on the licensor not to license other undertakings to exploit the licensed technology in the licensed territory;

Exclusive licence

(2) an obligation on the licensor not to exploit the licensed technology in the licensed territory himself;

---

[9] [1962] JO 204.

Protection of licensor

(3) an obligation on the licensee not to exploit the licensed technology in the territory of the licensor within the common market; *(Article 1(1)(4)–(6))*

Manufacture *(Article 1(1)(4))*

(4) an obligation on the licensee not to manufacture or use the licensed product, or use the licensed process, in territories within the common market which are licensed to other licensees;

Active sales

(5) an obligation on the licensee not to pursue an active policy of putting the licensed product on the market in the territories within the common market which are licensed to other licensees, and in particular not to engage in advertising specifically aimed at those territories or to establish any branch or maintain an distribution depot there;

Passive sales

(6) an obligation on the licensee not to put the licensed product on the market in the territories licensed to other licensees within the common market in response to unsolicited orders; *(Article 2(1)(14))*

Trade marks

(7) an obligation on the licensee to use only the licensor's trademark or get up to distinguish the licensed product during the term of the agreement, provided that the licensee is not prevented from identifying himself as the manufacturer of the licensed products; *(Articles 2(1)(11) and 5(1)(4))*

Use licence only

(8) an obligation on the licensee to limit his production of the licensed product to the quantities he requires in manufacturing his own products and to sell the licensed product only as an integral part of or a replacement part for his own products or otherwise in connection with the sale of his own products, provided that such quantities are freely determined by the licensee. *(Article 2(1)(13))*

Duration of territorial protection—pure patent licences (Recital 12)

2. Where the agreement is a pure patent licensing agreement, the exemption of the obligations referred to in paragraph 1 is granted only to the extent that and for as long as the licensed product is protected by parallel patents, in the territories respectively of the licensee (points (1), (2), (7) and (8)), the licensor (point (3)) and other licensees (points (4) and (5)). The exemption of the obligation referred to in point (6) of paragraph 1 is granted for a period not exceeding five years from the date when the licensed product is first put on the market within the common market by one of the licensees, to the extent that and for as long as, in these territories, this product is protected by parallel patents.

Duration of territorial protection—pure know-how licences (Recital 13)

3. Where the agreement is a pure know-how licensing agreement, the period for which the exemption of the obligations referred to in points (1) to (5) of paragraph 1 is granted may not exceed ten years from the date when the licensed product is first put on the market within the common market by one of the licensees.

The exemption of the obligation referred to in point (6) of paragraph 1 is granted for a period not exceeding five years from the date when the licensed product is first put on the market within the common market by one of the licensees.

The obligations referred to in points (7) and (8) of paragraph 1 are exempted during the lifetime of the agreement for as long as the know-how remains secret and substantial.

However, the exemption in paragraph 1 shall apply only where the parties have identified in any appropriate form the initial know-how and any subsequent improvements to it which become available to one party and are communicated to the other party pursuant to the terms of the agreement and to the purpose thereof, and only for as long as the know-how remains secret and substantial. *(Article 10(1)–(4))*

Duration of territorial restraints—mixed licences (Recital 16)

4. Where the agreement is a mixed patent and know-how licensing agreement, the exemption of the obligations referred to in points (1) to (5) of paragraph 1 shall apply in Member States in which the licensed technology is protected by necessary patents for as long as the licensed product is protected in those Member States by such patents if the duration of such protection exceeds the periods specified in paragraph 3.

The duration of the exemption provided in point (6) of paragraph 1 may not exceed the five-year period provided for in paragraphs 2 and 3.

However, such agreements qualify for the exemption referred to in paragraph 1 only for as long as the patents remain in force or to the extent that the know-how is identified and for as long as it remains secret and substantial, whichever period is the longer. *(Article 10(1)–(9))*

Obligations with more limited scope

5. The exemption provided for in paragraph 1 shall also apply where in a particular agreement the parties undertake obligations of the types referred to in that paragraph but with a more limited scope than is permitted by that paragraph.

Article 2

White list (Recital 18)

1. Article 1 shall apply notwithstanding the presence in particular of any of the following clauses, which are generally not restrictive of competition:

Confidentiality

(1) an obligation on the licensee not to divulge the know-how communicated by the licensor; the licensee may be held to this obligation after the agreement has expired;

No sub-licences

(2) an obligation on the licensee not to grant sublicences or assign the licence;

Post-term use ban (Recital 20)

(3) an obligation on the licensee not to exploit the licensed know-how or patents after termination of the agreement in so far and as long as the know-how is still secret or the patents are still in force;

Feed- and grant-back (Recital 20)

(4) an obligation on the licensee to grant to the licensor a licence in respect of his own improvements to or his new applications of the licensed technology, provided:

— that, in the case of severable improvements, such a licence is not exclusive, so that the licensee is free to use his own improvements or to license them to third parties, in so far as that does not involve disclosure of the know-how communicated by the licensor that is still secret,

— and that the licensor undertakes to grant an exclusive or non-exclusive licence of his own improvements to the licensee; *(Article 3(6))*

Minimum quality specifications

(5) an obligation on the licensee to observe minimum quality specifications, including technical specifications, for the licensed product or to procure goods or services from the licensor or from an undertaking designated by the licensor, in so far as these quality specifications, products or services are necessary for:

(a) a technically proper exploitation of the licensed technology; or

(b) ensuring that the product of the licensee conforms to the minimum quality specifications that are applicable to the licensor and other licensees;

and to allow the licensor to carry out related checks; *(Article 4(2)(a))*

Help enforce rights

(6) obligations:

(a) to inform the licensor of misappropriation of the know-how or of infringements of the licensed patents; or

(b) to take or to assist the licensor in taking legal action against such misappropriation or infringements; *(Article 4(2)(b))*

Royalties (Recital 21)

(7) an obligation on the licensee to continue paying the royalties:

(a) until the end of the agreement in the amounts, for the periods and according to the methods freely determined by the parties, in the event of the know-how becoming publicly known other than by action of the licensor, without prejudice to the payment of any additional damages in the event of the know-how becoming publicly known by the action of the licensee in breach of the agreement;

(b) over a period going beyond the duration of the licensed patents, in order to facilitate payment; *(Article 3)*

Field-of-use test (Recital 22)

(8) an obligation on the licensee to restrict his exploitation of the licensed technology to one or more technical fields of application covered by the licensed technology or to one or more product markets; *(Article 3(4))*

Minimum royalty and so on

(9) an obligation on the licensee to pay a minimum royalty or to produce a minimum quantity of the licensed product or to carry out a minimum number of operations exploiting the licensed technology; *(Articles 3(2) and 7(4))*

Most favoured licensee

(10) an obligation on the licensor to grant the licensee any more favourable terms that the licensor may grant to another undertaking after the agreement is entered into;

Mark product

(11) an obligation on the licensee to mark the licensed product with an indication of the licensor's name or of the licensed patent;

Not to construct facilities (Recital 24)

(12) an obligation on the licensee not to use the licensor's technology to construct facilities for third parties; this is without prejudice to the right of the licensee to increase the capacity of his facilities or to set up additional facilities for his own use on normal commercial terms, including the payment of additional royalties; *(Article 3(5))*

Second source (Recital 23)

(13) an obligation on the licensee to supply only a limited quantity of the licensed product to a particular customer, where the licence was granted so that the customer might have a second source of supply inside the licensed territory; this provision shall also apply where the customer is the licensee, and the licence which was granted in order to provide a second source of supply provides that the customer is himself to manufacture the licensed products or to have them manufactured by a subcontractor; *(Article 1(1)(8))*

Where no exhaustion (Recital 11)

(14) a reservation by the licensor of the right to exercise the rights conferred by a patent to oppose the exploitation of the technology by the licensee outside the licensed territory;

Rights if challenge to patent or know-how

(15) a reservation by the licensor of the right to terminate the agreement if the licensee contests the secret or substantial nature of the licensed know-how or challenges the validity of licensed patents within the common market belonging to the licensor or undertakings connected with him; *(Article 4(2)(b))*

Rights if alleged that patent not necessary (Recital 5)

(16) a reservation by the licensor of the right to terminate the licence agreement of a patent if the licensee raises the claim that such a patent is not necessary; *(Article 10(5))*

Licensee's best endeavours

(17) an obligation on the licensee to use his best endeavours to manufacture and market the licensed product; *(Articles 3(2) and 7(4))*

Licensee's rights if licence competes

(18) a reservation by the licensor of the right to terminate the exclusivity granted to the licensee and to stop licensing improvements to him when the licensee enters into competition within the common market with the licensor, with undertakings connected with the licensor or with other undertakings in respect of research and development, production, use or distribution of competing products, and to require the licensee to prove that the licensed know-how is not being used for the production of products and the provision of services other than those licensed. *(Article 3(2))*

Exemption

2. In the event that, because of particular circumstances, the clauses referred to in paragraph 1 fall within the scope of Article 85(1), they shall also be exempted even if they are not accompanied by any of the obligations exempted by Article 1.

Clauses of more limited scope

3. The exemption in paragraph 2 shall also apply where an agreement contains clauses of the types referred to in paragraph 1 but with a more limited scope than is permitted by that paragraph.

Article 3

Black list (Recitals 19 and 21)

Article 1 and Article 2(2) shall not apply where:

Price restrictions (Recital 24)

(1) one party is restricted in the determination of prices, components of prices or discounts for the licensed products;

No competition

(2) one party is restricted from competing within the common market with the other party, with undertakings connected with the other party or with other undertakings in respect of research and development, production, use or distribution of competing products without prejudice to the provisions of Article 2(1)(17) and (18); *(Article 2(1)(9), (17) and (18))*

Export restraints (Recitals 11 and 17)

(3) one or both of the parties are required without any objectively justified reason:

(a) to refuse to meet orders from users or resellers in their respective territories who would market products in other territories within the common market;

(b) to make it difficult for users or resellers to obtain the products from other resellers within the common market, and in particular to exercise intellectual property rights or take measures so as to prevent users or resellers from obtaining outside, or from putting on the market in the licensed territory products which have been lawfully put on the market within the common market by the licensor or with his consent; *(Articles 2(10)(14) and 7(3))*

or do so as a result of a concerted practice between them;

Customer restraints (Recital 23)

(4) the parties were already competing manufacturers before the grant of the licence and one of them is restricted, within the same technical field of use or within the same product market, as to the customers he may serve, in particular by being prohibited from supplying certain classes of user, employing certain forms of distribution or, with the aim of sharing customers, using certain types of packaging for the products, save as provided in Article 1(1)(7) and Article 2(1)(13); *(Article 2(1)(8))*

Quantity limits (Recital 24)

(5) the quantity of the licensed products one party may manufacture or sell or the number of operations exploiting the licensed technology he may carry out are subject to limitations, save as provided in Article (1)(8) and Article 2(1)(13);

Strong feed back (Recital 20)

(6) the licensee is obliged to assign in whole or in part to the licensor rights to improvements to or new applications of the licensed technology; *(Article 2(1)(4))*

Excessively long territorial restraints (Recital 14)

(7) the licensor is required, albeit in separate agreements or through automatic prolongation of the initial duration of the agreement by the inclusion of any new improvements, for a period exceeding that referred to in Article 1(2) and (3) not to license other undertakings to exploit the

licensed technology in the licensed territory, or a party is required for a period exceeding that referred to in Article 1(2) and (3) or Article 1(4) not to exploit the licensed technology in the territory of the other party or of other licensees.

Article 4

Opposition procedure (Recital 25)

1. The exemption provided for in Articles 1 and 2 shall also apply to agreements containing obligations restrictive of competition which are not covered by those Articles and do not fall within the scope of Article 3, on condition that the agreements in question are notified to the Commission in accordance with the provisions of Articles 1, 2 and 3 of Regulation (EC) No. 3385/94 and that the Commission does not oppose such exemption within a period of four months. *(Article 9)*

2. Paragraph 1 shall apply, in particular, where:

Quality specifications and ties

(a) the licensee is obliged at the time the agreement is entered into to accept quality specifications or further licences or to procure goods or services which are not necessary for a technically satisfactory exploitation of the licensed technology or for ensuring that the production of the licensee conforms to the quality standards that are respected by the licensor and other licensees; *(Article 2(1)(5))*

No challenge

(b) the licensee is prohibited from contesting the secrecy or the substantiality of the licensed know-how or from challenging the validity of patents licensed within the common market belonging to the licensor or undertakings connected with him. *(Article 2(1)(15))*

Four months to oppose

3. The period of four months referred to in paragraph 1 shall run from the date on which the notification takes effect in accordance with Article 4 of Regulation (EC) No 3385/94.

Transitional provision

4. The benefit of paragraphs 1 and 2 may be claimed for agreements notified before the entry into force of this Regulation by submitting a communication to the Commission referring expressly to this Article and to the notification. Paragraph 3 shall apply *mutatis mutandis*.

Rights and Member States

5. The Commission may oppose the exemption within a period of four months. It shall oppose exemption if it receives a request to do so from a Member State within two months of the transmission to the Member State of the notification referred to in paragraph 1 or of the communication referred to in paragraph 4. This request must be justified on the basis of considerations relating to the competition rules of the Treaty.

Withdraw the opposition

6. The Commission may withdraw the opposition to the exemption at any time. However, where the opposition was raised at the request of a Member State and this request is maintained, it may be withdrawn only after consultation of the Advisory Committee on Restrictive Practices and Dominant Positions.

Date of exemption

7. If the opposition is withdrawn because the undertakings concerned have shown that the conditions of Article 85(3) are satisfied, the exemption shall apply from the date of notification.

Date of exemption

8. If the opposition is withdrawn because the undertakings concerned have amended the agreement so that the conditions of Article 85(3) are satisfied, the exemption shall apply from the date on which the amendments take effect.

Notification may operate under Regulation 17

9. If the Commission opposes exemption and the opposition is not withdrawn, the effects of the notification shall be governed by the provisions of Regulation No 17.

Article 5

Exclusions (Recitals 6 and 8)

1. This Regulation shall not apply to:

Patent or know-how pool

(1) agreements between members of a patent or know-how pool which relate to the pooled technologies; *(Article 5(2)(2))*

Joint ventures between competitors

(2) licensing agreements between competing undertakings which hold interests in a joint venture, or between one of them and the joint venture, if the licensing agreements relate to the activities of the joint venture; *(Article 5(2)(1) and 5(3))*

Reciprocal rights between competitors

(3) agreements under which one party grants the other a patent and/or know-how licence and in exchange the other party, albeit in separate agreements or through connected undertakings, grants the first party a patent, trademark or know-how licence or exclusive sales rights, where the parties are competitors in relation to the products covered by those agreements; *Article 5(2)(2))*

Licences of other intellectual property rights (Recital 6)

(4) licensing agreements containing provisions relating to intellectual property rights other than patents which are not ancillary; *(Article 10(15))*

Sales licences (Recital 8)

(5) agreements entered into solely for the purpose of sale.

2. This Regulation shall nevertheless apply:

Joint ventures and market share

(1) to agreements to which paragraph 1(2) applies, under which a parent undertaking grants the joint venture a patent or know-how licence, provided that the licensed products and the other goods and services of the participating undertakings which are considered by users to be interchange-able or substitutable in view of their characteristics, price and intended use represent:

If distribution not joint

— in case of a licence limited to production, not more than 20%, and

If distribution joint

— in case of a licence covering production and distribution, not more than 10%;

of the market for the licensed products and all interchangeable or substitutable goods and services; *(Article 5(1)(2) and (2)(3))*

Where no territorial restraints

(2) to agreements to which paragraph 1(1) applies and to reciprocal licences within the meaning of paragraph 1(3), provided the parties are not subject to any territorial restriction within the common market with regard to the manufacture, use or putting on the market of the licensed products or to the use of the licensed or pooled technologies. *(Article 5(1)(1) and (3))*

Joint venture—market share—marginal relief

3. This Regulation shall continue to apply where, for two consecutive financial years, the market shares in paragraph 2(1) are not exceeded by more than one-tenth; where that limit is exceeded, this Regulation shall continue to apply for a period of six months from the end of the year in which the limit was exceeded.

Article 6

Inclusions (Recital 9)

This Regulation shall also apply to:

Sub-licences

(1) agreements where the licensor is not the holder of the know-how or the patentee, but is authorized by the holder or the patentee to grant a licence;

Assignments

(2) assignments of know-how, patents or both where the risk associated with exploitation remains with the assignor, in particular where the sum

payable in consideration of the assignment is dependent on the turnover obtained by the assignee in respect of products made using the know-how or the patents, the quantity of such products manufactured or the number of operations carried out employing the know-how or the patents;

Connected undertakings

(3) licensing agreements in which the rights or obligations of the licensor or the licensee are assumed by undertakings connected with them. *(Article 10(14))*

Article 7

Withdrawal of exemption (Recital 26)

The Commission may withdraw the benefit of this Regulation, pursuant to Article 7 of Regulation No 19/65/EEC, where it finds in a particular case that an agreement exempted by this Regulation nevertheless has certain effects which are incompatible with the conditions laid down in Article 85(3) of the Treaty, and in particular where:

No effective inter-brand competition

(1) the effect of the agreement is to prevent the licensed products from being exposed to effective competition in the licensed territory from identical goods or services or from goods or services considered by users as interchangeable or substitutable in view of their characteristics, price and intended use, which may in particular occur where the licensee's market share exceeds 40%;

Licensee refuses passive sales

(2) without prejudice to Article 1(1)(6), the licensee refuses, without any objectively justified reason, to meet unsolicited orders from users or resellers in the territory of other licensees;

Parties hinder parallel trade

(3) the parties:
- (a) without any objectively justified reason, refuse to meet orders from users or resellers in their respective territories who would market the products in other territories within the common market; or
- (b) make it difficult for users or resellers to obtain the products from other resellers within the common market, and in particular where they exercise intellectual property rights or take measures so as to prevent resellers or users from obtaining outside, or from putting on the market in the licensed territory products which have been lawfully put on the market within the common market by the licensor or with his consent; *(Article 3(3))*

Competing manufacturers—minimum quantities—best endeavours

(4) the parties were competing manufacturers at the date of the grant of the licence and obligations on the licensee to produce a minimum quantity or

to use his best endeavours as referred to in Article 2(1), (9) and (17) respectively have the effect of preventing the licensee from using competing technologies.

Article 8

Inclusions

1. For purposes of this Regulation:

   (a) patent applications;
   (b) utility models;
   (c) applications for registration of utility models;
   (d) topographies of semiconductor products;
   (e) *certificats d'utilité* and *certificats d'addition* under French law;
   (f) applications for *certificats d'utilité* and *certificats d'addition* under French law;
   (g) supplementary protection certificates for medicinal products or other products for which such supplementary protection certificates may be obtained;
   (h) plant breeder's certificates,
       shall be deemed to be patents.

Applications within one year

2. This Regulation shall also apply to agreements relating to the exploitation of an invention if an application within the meaning of paragraph 1 is made in respect of the invention for a licensed territory after the date when the agreements were entered into but within the time-limits set by the national law or the international convention to be applied.

Initial duration prolonged (Recital 14)

3. This Regulation shall furthermore apply to pure patent or know-how licensing agreements or to mixed agreements whose initial duration is automatically prolonged by the inclusion of any new improvements, whether patented or not, communicated by the licensor, provided that the licensee has the right to refuse such improvements or each party has the right to terminate the agreement at the expiry of the initial term of an agreement and at least every three years thereafter.

Article 9

Confidentiality of notification

1. Information acquired pursuant to Article 4 shall be used only for the purposes of this Regulation.

2. The Commission and the authorities of the Member States, their officials and other servants shall not disclose information acquired by them pursuant to this Regulation of the kind covered by the obligation of professional secrecy.

3. The provisions of paragraphs 1 and 2 shall not prevent publication of general information or surveys which do not contain information relating to particular undertakings or associations of undertakings.

Article 10

Definitions

For purposes of this Regulation:

'Know-how'

(1) 'know-how' means a body of technical information that is secret, substantial and identified in any appropriate form;

'Secret'

(2) 'secret' means that the know-how package as a body or in the precise configuration and assembly of its components is not generally known or easily accessible, so that part of its value consists in the lead which the licensee gains when it is communicated to him; it is not limited to the narrow sense that each individual component of the know-how should be totally unknown or unobtainable outside the licensor's business;

'Substantial'

(3) 'substantial' means that the know-how includes information which must be useful, i.e. can reasonably be expected at the date of conclusion of the agreement to be capable of improving the competitive position of the licensee, for example by helping him to enter a new market or giving him an advantage in competition with other manufacturers or providers of services who do not have access to the licensed secret know-how or other comparable secret know-how;

'Identified'

(4) 'identified' means that the know-how is described or recorded in such a manner as to make it possible to verify that it satisfies the criteria of secrecy and substantiality and to ensure that the licensee is not unduly restricted in his exploitation of his own technology, to be identified the know-how can either be set out in the licence agreement or in a separate document or recorded in any other appropriate form at the latest when the know-how is transferred or shortly thereafter, provided that the separate document or other record can be made available if the need arises;

'Necessary patents'

(5) 'necessary patents' are patents where a licence under the patent is necessary for the putting into effect of the licensed technology in so far as, in the absence of such a licence, the realization of the licensed technology would not be possible or would be possible only to a lesser extent or in more difficult or costly conditions. Such patents must therefore be of technical, legal or economic interest to the licensee;

'Licensing agreement'

(6) 'licensing agreement' means pure patent licensing agreements and pure know-how licensing agreements as well as mixed patent and know-how licensing agreements;

'Licensed technology'

(7) 'licensed technology' means the initial manufacturing know-how or the necessary product and process patents, or both, existing at the time the first licensing agreement is concluded, and improvements subsequently made to the know-how or patents, irrespective of whether and to what extent they are exploited by the parties or by other licensees;

'The licensed products'

(8) 'the licensed products' are goods or services the production or provision of which requires the use of the licensed technology;

'The licensee's market share'

(9) 'the licensee's market share' means the proportion which the licensed products and other goods or services provided by the licensee, which are considered by users to be interchangeable or substitutable for the licensed products in view of their characteristics, price and intended use, represent the entire market for the licensed products and all other interchangeable or substitutable goods and services in the common market or a substantial part of it; *(Contrast Article 10(17))*

'Exploitation'

(10) 'exploitation' refers to any use of the licensed technology in particular in the production, active or passive sales in a territory even if not coupled with manufacture in that territory, or leasing of the licensed products;

'The licensed territory'

(11) 'the licensed territory' is the territory covering all or at least part of the common market where the licensee is entitled to exploit the licensed technology;

'Territory of the licensor'

(12) 'territory of the licensor' means territories in which the licensor has not granted any licences for patents and/or know-how covered by the licensing agreement;

'Parallel patents'

(13) 'parallel patents' means patents which, in spite of the divergences which remain in the absence of any unification of national rules concerning industrial property, protect the same invention in various Member States;

'Connected undertakings'

(14) 'connected undertakings' means:

    (a) undertakings in which a party to the agreement, directly or indirectly:

        — owns more than half the capital or business assets, or

        — has the power to exercise more than half the voting rights, or

        — has the power to appoint more than half the members of the

supervisory board, board of directors or bodies legally representing the undertaking, or
— has the right to manage the affairs of the undertaking;

(b) undertakings which, directly or indirectly, have in or over a party to the agreement the rights or powers listed in (a);

(c) undertakings in which an undertaking referred to in (b), directly or indirectly, has the rights or powers listed in (a);

(d) undertakings in which the parties to the agreement or undertakings connected with them jointly have the rights or powers listed in (a): such jointly controlled undertakings are considered to be connected with each of the parties to the agreement;

'Ancillary provisions'

(15) 'ancillary provisions' are provisions relating to the exploitation of intellectual property rights other than patents, which contain no obligations restrictive of competition other than those also attached to the licensed know-how or patents and exempted under this Regulation;

'Obligation'

(16) 'obligation' means both contractual obligation and a concerted practice;

'Competing manufacturers'—'competing products'

(17) 'competing manufacturers' or manufacturers of 'competing products' means manufacturers who sell products which, in view of their characteristics, price and intended use, are considered by users to be interchangeable or substitutable for the licensed products.

Article 11

Transitional provisions (Recital 3)

Repeal of know-how regulation

1. Regulation (EEC) No 556/89 is hereby repealed with effect from 1 April 1996.

Extension of patent regulation

2. Regulation (EEC) No 2349/84 shall continue to apply until 31 March 1996.

Old block exempted agreements

3. The prohibition in Article 85(1) of the Treaty shall not apply to agreements in force on 31 March 1996 which fulfil the exemption requirements laid down by Regulation (EEC) No 2349/84 or (EEC) No 556/89.

Article 12

Reassessment of this regulation

1. The Commission shall undertake regular assessments of the application of this Regulation, and in particular of the opposition procedure provided for in Article 4.

2. The Commission shall draw up a report on the operation of this Regulation before the end of the fourth year following its entry into force and shall, on that basis, assess whether any adaptation of the Regulation is desirable.

Article 13

Entry into force

This Regulation shall enter into force on 1 April 1996.

It shall apply until 31 March 2006.

Article 11(2) of this Regulation shall, however, enter into force on 1 January 1996.

This Regulation shall be binding in its entirety and directly applicable in all Member States.

Done at Brussels, 31 January 1996.

For the Commission
Karel VAN MIERT
Member of the Commission